International Conference on Autonomous Ships 2020

London, United Kingdom
17 - 18 June 2020

ISBN: 978-1-7138-8295-4

Printed from e-media with permission by:

Curran Associates, Inc.
57 Morehouse Lane
Red Hook, NY 12571

Some format issues inherent in the e-media version may also appear in this print version.

The Royal Institution of Naval Architects
8-9 Northumberland Street
London, WC2N 5DA
United Kingdom

Phone: 020 7235 4622
Fax: 020 7259 5912

publications@rina.org.uk

Additional copies of this publication are available from:

Curran Associates, Inc.
57 Morehouse Lane
Red Hook, NY 12571 USA
Phone: 845-758-0400
Fax: 845-758-2633
Email: curran@proceedings.com
Web: www.proceedings.com

International Conference

Autonomous Ships

17th-18th June 2020, London, UK

PAPERS

Sponsored by:

LLOYD'S MARITIME ACADEMY

THE ROYAL INSTITUTION OF NAVAL ARCHITECTS
8-9 Northumberland Street, London, WC2N 5DA, Telephone: +44 (0)20 7235 4622

The Royal Institution of
Naval Architects

INTERNATIONAL CONFERENCE

Autonomous Ships

17th – 18th June 2020, London, UK

Papers published in these conference Proceedings are peer reviewed.

THE ROYAL INSTITUTION OF NAVAL ARCHITECTS
8-9 Northumberland Street
London WC2N 5DA

Telephone: 020 7235 4622
Fax: 020 7259 5912

CONTENTS

TOWARDS AUTONOMOUS SHIPS – FLAG STATE INVOLVEMENT AND REGULATORY ASPECTS

C. Balls, Cayman Registry, Cayman Islands

SUMMARY

IMO interim guidelines for MASS trials – MSC.1/Circ.1604 –makes it clear that Flag States have prime responsibility for the safe operation of MASS (Maritime Autonomous Surface Ships) Jurisdiction of vessels is determined by the UN convention on the Law of the Seas (UNCLOS). The four main pillars of Maritime Regulation are considered to be; SOLAS; MARPOL; STCW; MLC. With the other most relevant conventions being; ILL; COLREG; BWM. As well as the Liability and Compensation Conventions especially CLC, Bunker Oil, PAL and LLMC. Search and Rescue also needs to be considered. There already exists provision for Alternative Design, Exemptions and equivalences in various conventions. The latest version of the Large Yacht Code – The Red Ensign Group Yacht Code introduced last year has developed further these concepts for Large Yachts. Smaller vessels (especially under 24m L) are regulated to a lesser extent, although codes, such as the Caribbean Small Commercial Vessel Code, are intended to provide standards for internationally trading small vessels and already mention some of the considerations for MASS. Finally, the paper will look in more detail at, some of the issues related to navigation and moving towards conditionally and periodically unmanned bridges - B0

1. INTRODUCTION

1.1 Shipping is a highly regulated Industry with a vessels Flag State having responsibility for ensuring that vessels registered under its flag comply with the numerous international regulations, however, it is still possible to progress introduction of new technology. IMO interim guidelines for MASS trials – MSC.1/Circ.1604 – makes it clear that Flag States have prime responsibility for the safe operation of MASS

1.2 This paper will take some of the relevant conventions and describe how Autonomous Ship related developments have occurred and future possibilities even within the constraints of current regulation. An autonomous ship is not necessarily unmanned but an unmanned ship must be autonomous to a high degree.

1.3 The four main pillars of Maritime Regulation are considered ; The International Convention for the Safety of Life at Sea (SOLAS); The International Convention for the Prevention of Pollution from Ships (MARPOL); International Convention on Standards of Training Certification and Watch-keeping for Seafarers (STCW); and the Maritime Labour Convention (MLC).

The first three being Conventions made under IMO and the fourth under the ILO

1.4 The following IMO conventions are also relevant; The International Convention on Load-lines (ILL); The); The Convention on the International Regulations for the Prevention of Collisions at Sea (COLREG); The International Convention for the Control and Management of Ships' Ballast Water and Sediments (BWM); The International Convention on Civil Liability for Oil Pollution Damage (CLC); The Athens Convention relating to the Carriage of Passengers and their Luggage by Sea (PAL); Convention on Limitation of Liability for Maritime Claims (LLMC); International Convention on Civil Liability for Bunker Oil Pollution Damage; Nairobi International Convention on the Removal of Wrecks (WRC) and; The International Convention on Maritime Search and Rescue (SAR).

1.5 The paper will also look in particular at the moves toward more automation in Navigation Functions as well as use of relevant vessel Codes especially for smaller vessels which is where much of autonomous ship development is starting.

2. SOLAS

This convention is the main one for regulating the safety aspects of the design construction and operation of ships.

2.1 GENERAL

This chapter allows for exemptions for ships which embody features of a novel kind from any of the provisions of chapters II-1, II-2, III and IV of the regulations the application of which might seriously impede research into the development of such features and their incorporation in ships engaged on international voyages. Any such ship shall, however, comply with safety requirements which, in the opinion of that Administration, are adequate for the service for which it is intended and are such as to ensure the overall safety of the ship and which are acceptable to the Governments of the States to be visited by the ship This does effectively restrict these exemptions to a small number of vessels but can be used to assist with the development of new technology.
The equivalents provision is probably the most appropriate method of enabling development of autonomous ships IMO issues guidelines for the appropriate way of assessing equivalence such as;

MSC.1/Circ.1212 Guidelines on alternative design and arrangements for SOLAS Chapters II-1 and III. These particular guidelines serve to outline the methodology for the engineering analysis required by SOLAS regulations II-1/55 and III/38 on alternative design and arrangements, applying to a specific engineering or life-saving system, design or arrangements for which the approval of an alternative design deviating from the prescriptive requirements of SOLAS chapters II-1 and III is sought.

2.2 CONSTRUCTION

This area allows some flexibility in the application section to allow exemption for vessels which, in the course of their voyage, do not proceed more than 20 miles from the nearest land if the Administration considers that the sheltered nature and conditions of the voyage are such as to render the application of any specific requirements of this chapter unreasonable or unnecessary, The move towards Goal based methodologies rather than completely prescriptive ones also helps with the increasing use of autonomous systems. Machinery installations and electrical installations respectively have requirements mainly based on relatively traditional arrangements although again there has been sufficient flexibility to allow the development of such things as azimuthing propulsion systems and work continues which in turn enable utilization for alternative fuel systems to improve environmental aspects of shipping. Automated systems are being utilized to improve monitoring of machinery and improve operational efficiency as well improved communication systems which allow for remote monitoring and diagnostic work to be undertaken, much of this is in addition to the prescriptive requirements thus there is more flexibility in development.

There have been trends to make much more technical information available to Officers on the Navigation Bridge but of course the possibilities of B0 type operations could well mean that the opposite occurs and many Navigation functions are conducted from the engine control room.

Periodically unattended machinery spaces, which were one of the earlier additions to the SOLAS regulations to take into account automation developments, enabled crew sizes to be reduced which was largely seem by the Industry as beneficial from the operating costs point of view. As technology develops, we could soon be seeing periodically unattended bridges being allowed.

2.3 LIFESAVING APPLIANCES AND ARRANGEMENTS

The probable ever decreasing crew numbers will inevitably lead to changes in the provision of lifesaving appliances and there will be a need to ensure that they are appropriate for the maximum numbers of persons envisaged to be on board for example if technicians have to be embarked to make and verify repairs.

2.4 RADIO COMMUNICATIONS

There have been seen significant changes over the years as technology has changed and improved the ability to transmit vast amounts of data around the world and the chapter relies heavily on performance standards which are frequently reviewed and work is coordinated with other relevant organizations to enable the co-existence of systems needing to utilize radio bandwidth without interference. Ensuring robust and secure communication systems is of course one of the key factors in the safe development of autonomous systems.

2.5 CARRIAGE OF CARGOES

This is an area where automated monitoring will need further development and of course there may well be varying levels of public acceptance of the types of vessel that are permitted to operate without what is traditionally called a crew, for example in particular Passenger ships and Tankers/ Gas Carriers. Although of course many other ship types carry dangerous goods but less obviously.

2.6 MANAGEMENT FOR THE SAFE OPERATION OF SHIPS

This part of SOLAS covers operational aspects and brings into force the ISM Code.. Section 5 of the code on Masters Responsibility and Authority of course is an area where some consider developments in communications have undermined the Masters authority and of course if we get to the stage of unmanned ships over 500 GT who is the master and where are they located? Section 9 covering reports and analysis of non-conformances and hazardous occurrences has benefited from the ever-increasing amount of data available including better voyage data recording which has been enabled by technical developments giving possibilities of further enhancement of analysis possibilities. Section 10 on Maintenance of the ship and equipment also benefits from technological developments. Because the code only gives a very broad overview of what is required rather than prescriptive requirements there is little restriction on development of autonomous technology and developing predictive maintenance programs in order to prevent breakdowns and optimize the carriage of spare parts and effective utilization of skilled personnel.

2.7 MARITIME SECURITY.

This covers the ISPS Code including requirements for ship security alert systems. Shipboard security systems have benefited from modern technology in most areas with better physical security by sophisticated door locking and control systems, CCTV and other monitoring systems will be part of autonomous technology ,as of course is the development of secure communication systems so vital for combating the security threats posed. The provision in regulation 12 for equivalent security arrangements aids developments of new solutions with respect to autonomy.

2.8 HIGH-SPEED CRAFT

This chapter enables the use of the High Speed craft code. The latest code applies to new builds which are either passenger craft which do not proceed in the course of their voyage more than four hours at 90% of maximum speed from a place of refuge or cargo craft of 500 gross tonnage and upwards which do not proceed in the course of their voyage more than 8 hours at 90% of maximum speed from a place of refuge when fully laden. These vessels can benefit from autonomous technology and are fitted with modern systems to allow for their safe operation with the speeds achieved necessitating the development of much improved bridge ergonomics and also of course many differences from traditional working arrangements with the engineer often based on the bridge. There is an allowance for equivalences which helps give flexibility for development of systems in the code. Chapter 1 1.11.1 of the code states that where it requires that a particular fitting, material, appliance or apparatus, or type thereof, shall be fitted or carried in a craft, or that any particular provision shall be made, the Administration may allow any other fitting, material, appliance or apparatus, or type thereof, to be fitted or carried, or any other provision to be made in the craft, if it is satisfied by trial thereof or otherwise that this is at least as effective as that required by the Code. Whilst these vessels have already made use of the technology related to development of autonomous vessels they are likely to remain with manned bridges for some time.

3. MARPOL

3.1 ANNEX 1 REGULATIONS FOR PREVENTION OF POLLUTION BY OIL

Although in general there are few exemption provisions in MARPOL when compared to SOLAS this annex does contain exemption provisions at regulation 3 which states that any ship such as hydrofoil, air-cushion vehicle, near-surface craft and submarine craft etc. whose constructional features are such as to render the application of any of the provisions of chapters 3 and 4 of this Annex relating to construction and equipment unreasonable or impracticable may be exempted by the Administration from such provisions, provided that the construction and equipment of that ship provides equivalent protection against pollution by oil, having regard to the service for which it is intended. This is useful for permitting the development of such ships using alternative means to those mainly prescriptive ones in the annex.

This convention also contains unified interpretations which give more specific detail of various issues and these are referred to in regulations 3.4 and 3.5 which allow for waivers to be granted to certain vessels on certain voyages. Regulation 5 allows any fitting, material, appliance or apparatus to be fitted in a ship as an alternative to that required by the Annex if such fitting, material, appliance or apparatus is at least as effective as that required by this Annex. However this authority does not extend to substitution of operational methods to effect the control of discharge of oil as equivalent to those design and construction features which are prescribed by regulations in this Annex.

The layout and equipment is reasonably tightly controlled and technology is increasingly useful for ensuring compliance by helping the operators to determine vessel condition especially with respect to tank contents and likely outflows in event of damage as well as improving equipment monitoring for alarm conditions and shutdowns. The use of electronic log books and records is becoming more widely accepted for this and other annexes of MARPOL, these systems of course lend themselves far better to autonomous operations.

3.2 ANNEX II REGULATIONS FOR PREVENTION OF POLLUTION BY NOXIOUS LIQUID SUBSTANCES IN BULK

The Equivalence provisions of Regulation 5 are of more relevance to autonomy and the wording is similar to that in regulation 5 of Annex 1.

3.3 ANNEX III - REGULATIONS FOR THE PREVENTION OF POLLUTION BY HARMFUL SUBSTANCES CARRIED BY SEA IN PACKAGED FORM

This applies to all ship types and there are no exemptions or equivalences. Technological advances do of course help to improve monitoring of carriage and ensuring optimum solutions as well as helping in document transmission and tracking.

3.4 ANNEX IV REGULATIONS FOR THE PREVENTION OF POLLUTION BY SEWAGE FROM SHIPS

Discharge criteria also require proper monitoring again giving scope for better automation and accurate monitoring. Reducing numbers of seafarers on board will of course lead to less sewage production

3.5 ANNEX V - REGULATIONS FOR THE PREVENTION OF POLLUTION BY GARBAGE FROM SHIPS

In terms of development of technology this is similar to annex IV with no exemptions or equivalences and technological advances helping in many areas such as reducing garbage production and then optimizing the treatment of garbage by compacting or incinerating to give most effective disposal solutions.

3.6 ANNEX VI - REGULATIONS FOR THE PREVENTION OF AIR POLLUTION FROM SHIPS

This annex does have exemptions and equivalents provisions

Regulation 3 allows for exemption from specific provisions for the purpose of allowing trials for the development of ship emission reduction and control technologies and engine design programs under certain provisions. This is helpful for the development of solutions related to autonomy. There is also the option of exempting certain emissions from sea-bed mineral activities which also of course currently utilize remotely operated and autonomous vessels as with hydrographic activities.

Regulation 4 allows any fitting, material, appliance or apparatus to be fitted in a ship or other procedures, alternative fuel oils, or compliance methods used as an alternative to that required by the Annex if such fitting, material, appliance or apparatus or other procedures, alternative fuel oils, or compliance methods are at least as effective in terms of emissions reductions as that required by the Annex. The whole area of improving energy efficiency of ships is greatly aided by improving technology and automation to optimize performance. Ever more stringent targets are being set for emissions both at an individual ship level and also from the overall supply chain aspects. Ensuring and enforcing compliance with world-wide regulation will also utilize new technology to monitor compliance There will be an ever-increasing demand for accurate sensors and complex analysis of data.

4. STCW

4.1 CHAPTER 1 GENERAL PROVISIONS

This chapter contains definitions and clarifications. It covers the issuance and acceptance of certification and endorsements and also covers the principles of near coastal voyages. Control and National provisions are also addressed. It goes on to cover Training and Assessment, communication of information, quality and medical standards. Recognition and revalidation of certificates also comes under this chapter. The section on use of simulators is probably the one where developing technology comes to the fore as data received from real situations with vessels is vital to enable the necessary modelling for simulators. Another section covers trials which again are vital to allow new systems to be tested. Finally the responsibilities of companies are outlined. Increasing developments of technology and moves towards autonomy will need development of both the existing requirements and amendments to cope with changes necessitated by increasing autonomy and changing roles of personnel.

4.2 CHAPTER II MASTER AND DECK DEPARTMENT

This chapter covers standards of competence, on board training. It is divided into the following functions; Navigation, cargo handling and stowage, **Controlling the operation of the ship and care for persons on board,** all at both operational and management levels. This chapter has been updated to cover technological developments in many areas related to navigation such as ARPA and ECDIS whilst still maintains traditional skills such as use of Sextants. If the location of some or all of these functions moves then amendments will be necessary to enable practical application.

4.3 CHAPTER III ENGINE DEPARTMENT

This chapter covers standards of competence, on board training. It is divided into the following functions: Marine Engineering, Maintenance and repair, **controlling the operation of the ship and care for persons on board.** Again all at both operational and management levels. Technological developments have resulted in updating these requirements to cover such aspects as automation and also the introduction of electro technical requirements to reflect the growing importance of this role which has significant importance in ensuring the developing smart systems are properly supported. Updating similar to chapter II will be needed.

4.4 CHAPTER IV RADIO OPERATORS

This covers the minimum standard of competence for GMDSS radio operators at the operational level. The widespread removal of dedicated Radio Officers is a good example of how the roles of seafarers have changed with the onset of improved technological solutions. Again the possible relocation off the vessel will need some changes.

4.5 CHAPTER V SPECIAL TRAINING REQUIREMENTS FOR PERSONNEL ON CERTAIN TYPES OF SHIP

This chapter covers additional training requirements due to the complexity of certain ship types and has increased in size as new types are developed. It presently covers; minimum requirements for the training and qualifications of masters, officers and ratings on oil and chemical tankers *liquefied gas tankers. Also* minimum requirements for the training and qualification of masters, officers, ratings and other personnel on passenger ships covering such areas as crowd management, crisis management and human behavior as well as passenger safety, cargo safety and hull integrity training for certain personnel. Again all these requirements will need further amendments in the light of developing technology.

4.6 CHAPTER VI EMERGENCY, OCCUPATIONAL SAFETY, SECURITY, MEDICAL CARE AND SURVIVAL FUNCTIONS

This chapter requires that all persons employed or engaged on a seagoing ship, other than passengers, shall receive approved familiarization training in personal survival techniques or receive sufficient information and instruction, it goes on to specify appropriate approved basic training or instruction in: personal survival techniques, fire prevention and fire fighting, elementary first aid and personal safety and social responsibilities. It goes on to cover proficiency in survival craft, rescue boats and fast rescue boats, advanced fire-fighting, medical first aid and medical care. It also covers security-related training and instruction for all seafarers. Changing crew numbers and functions will probably led to changes in this chapter.

4.7 CHAPTER VII ALTERNATIVE CERTIFICATION

This chapter allows for alternatives with the stipulation that no alternative certification system shall be implemented unless it ensures a degree of safety at sea and has a preventive effect as regards pollution at least equivalent to that provided by the other chapters and it shall provide for the interchangeability of certificates with those issued under the other chapters. This provision does allow for certification systems to keep up with technological developments where different skill sets may get combined in a non-traditional manner which provides opportunities for effective introduction of autonomy although it does state that the principles of the regulation shall ensure that the competency of both deck and engineer officers is maintained, which could be interpreted as being a bit restrictive.

4.8 CHAPTER VIII WATCH KEEPING

This chapter covers fitness for duty and watch keeping arrangements with the principles to be observed. Here there continues to be research into the best ways to combat fatigue and ensure that working arrangements are such as to optimize seafarer performance. The research into this involves many sophisticated tools. The effective monitoring has seen electronic technology being introduced to ensure compliance with regulations which based on current knowledge have been developed to minimize the effects of fatigue.
The subject of keeping a proper watch and lookout is one that presents challenges to the development of more automated vessels especially autonomous ones and of course if vessels only need to be manned for certain proportion of their voyage the traditional thinking of watch-keeping arrangements will need to be revisited to ensure those persons on board are fit for their duties.

5. COLREGS

Rule 1 (e) of this convention allows a vessel of special construction or purpose to deviate from the provisions of any of these Rules with respect to the number, position, range or arc of visibility of lights or shapes, as well as to the disposition and characteristics of sound signaling appliances provided arrangements are as close as possible to compliant. This gives flexibility for developing ship designs. Rule 5 states that every vessel shall at all times maintain a proper look-out by sight and hearing as well as by all available means appropriate in the prevailing circumstances and conditions so as to make a full appraisal of the situation and of the risk of collision.– but maybe with appropriate technology could also be done by an operator ashore or even eventually autonomously Compliance with Rules 7 (risk of collision) and Rule 8 (Action to avoid collision) has been aided by the development of radar systems and computer developments enable optimal solutions to avoiding collisions to be found which should reduce the deviations from optimum passage speeds and courses, but care is needed in ensuring suitable parameters are always applied. Of course care also need to be taken to ensure that all ships do not try to follow the same optimum tracks.

6. LOADLINE

This convention has similar provisions to SOLAS which enables the development of new technology within the guidelines of the existing convention. These are again in the form of exemption and equivalence provisions. The areas of the convention most likely to be affected by autonomy cover such aspects as; protection of openings, guard rails and means of access to crew's quarters.

7. BWM

This convention also needs to be considered as whilst many ballast water treatment systems are quite automated there is usually still need for human intervention for such things as filter changes or lamp changes in ultra violet treatment systems so her may be need for more backup systems if humans are removed.

8. MLC

8.1 The tripartite (Government, Ship owners and Seafarers Unions) arrangement for consultation regarding changes in this Convention may be seen, by some, as having potential for slowing development due to concerns on job losses etc., however, it does benefit operations by aiding the interface between those involved in ship and shore operations. There is provision for Substantial Equivalence as described in Articles VI VII and XIII.

8.2 Progress towards unmanned ships will raise new challenges, such as those presented by ships which are autonomous for part of the voyage and then manned for arrivals and departures. This can potentially give rise to Health and Safety (Title 4-Regulation 4.3) concerns regarding embarking / disembarking persons for events such as coastal passages and port entry; here technology needs further development, such as seen in personnel transfer solutions seen in the Offshore Industry. The advent of vessels which are sometimes unmanned may also present some interesting interpretations of Employment Contract requirements (Title 2 – Regulation 2.1). Crew accommodation requirements (Title 3 – Regulation 3.1) are very basic, although quite prescriptive, with little opportunity for exemption These requirements should not unduly impede development, as they are minimum standards and few Seafarers would be happy working at sea with poorer conditions. Technology, much of it automated, can improve Seafarer comfort by reducing noise and vibration s well as stabilizing vessel movement

8.3 Un-manned ships takes many aspects of autonomy to the extreme and indeed if all ships were totally un-manned there would be no need for the MLC convention.

8.4 One possible downside of taking people off ships is that currently there is at least an element of Seafarer self-preservation which helps to preserve assets by preventing casualties. Of course, technology is of great assistance to the current day seafarer in preventing casualties albeit with the various challenges of potential information overload and undetected equipment errors.

9. THE LIABILITY AND COMPENSATION CONVENTIONS

These conventions cover such things as; Oil Pollution, Carriage of Passengers and their Luggage by Sea, Limitation of Liability for Maritime Claims, Civil Liability for Bunker Oil Pollution Damage; and Removal of Wrecks. The conventions and their enabling legislation re of course based on the assumption that vessels are manned, if this is not the case then some of the terminology will probably need modification to close any potential legal loopholes. There are bound to be errors in autonomous system programming and where the liability lies in these cases will need to be determined to establish who needs appropriate insurance coverage.

10. THE INTERNATIONAL CONVENTION ON MARITIME SEARCH AND RESCUE.

The use of drones for SAR is already being developed as are various devices to enable remote casualty recovery. The development of the sensors required by autonomous ships could aid the safe conduct of search and rescue by reducing the need for humans to take risks to assist others.

11. EXAMPLES OF POSSIBLE INTRODUCTION OF AUTONOMY TO VESSELS OPERATING UNDER EXISTING VESSEL CODES

11.1 The Code of Safety for Small Commercial Vessels operating in the Caribbean has the following text in Chapter 1 section 34 with respect to autonomous and unmanned vessels;

- The control of these vessels rests with persons not located on board, with the potential for fully autonomous vessels to conduct voyages entirely with no human intervention, whilst unmanned vessels may be controlled, on their voyages, from a remote location;
- These types of vessels will be subject to special consideration and will require exemptions from many provisions of the Code primarily because there are no seafarers on board to operate the vessel and its equipment. Appropriate equivalences shall be determined dependent upon the nature of the operations;
- Operations involving these vessels shall be fully risk assessed with consideration being given to such aspects as interrelation with other vessels and possible safe human intervention requirements;
- The operation of these vessels will require the issuance of Notices to Mariners and the necessary information shall be provided to the relevant authorities to enable the appropriate promulgation of such notices;
- Whilst these vessels are not subject to manning requirements, those responsible for their operation should be appropriately qualified, trained and considered competent to control such vessels; and
- Special consideration will be required where the operators are located outside of the territory in which the vessel is operating.

The Code is being reviewed this year and will update this section to reflect developments in MASS and probably included the latest relevant definitions.

11.2 The Red Ensign Group Yacht Code refers to the guidelines on alternative design and arrangements for SOLAS chapters ii-1 and iii (msc.1/circ.1212) and SOLAS chapters ii-2 (msc.1/circ.1002) in part A Section 1.9 entitled alternative design and arrangements with the purpose of providing a methodology for alternative design and arrangements which are limited by the extent to which international conventions allow. Only chapters containing provisions for alternative design and arrangements (7,8,9,10,13,14,15) may utilize such provisions.

11.3 The Red Ensign Group Yacht Code guidelines on alternative design and arrangements include the following aspects.

- design and arrangements may deviate from the requirements of Part A of the code, provided that the alternative design and arrangements meet the intent of the requirements concerned, the chapter's objectives and functional requirements where provided, and provide an equivalent level of safety to the chapter concerned. Alternative design and arrangements shall be based on the underlying convention requirements and follow any administrations alternative design and arrangement procedures.

- When alternative design or arrangements deviate from the prescriptive requirements of Part A, an engineering analysis, evaluation and approval of the design and arrangements shall be carried out in accordance with this section.
- The engineering analysis shall be prepared and submitted to the administration, based on the guidelines and shall include, as a minimum, the elements listed under 'engineering analysis elements' in the relevant chapter.
- The engineering analysis shall be evaluated and approved by the administration, considering the guidelines. A copy of the documentation, as approved by the administration, indicating that the alternative design and arrangements comply with this section, shall be carried onboard the ship.
- The administration shall communicate to the IMO pertinent information concerning alternative design and arrangements approved by them for circulation to all contracting governments.
- Re-evaluation due to change of conditions if the alternative design and arrangements are changed, the engineering analysis shall be carried out under the changed condition and the re-evaluation shall be approved by the administration.

11.4 The Red Ensign Group Yacht Code engineering analysis requirements typically involve the following elements:

- Determination of the ship type and relevant arrangements;
- Identification of the prescriptive requirement(s);
- Appliances and arrangements will not comply;
- Identification of the reason the proposed design will not meet the prescriptive requirements supported by compliance with other recognized engineering or industry standards;
- Determination of the performance criteria for the ship and appliances or arrangements concerned addressed by the relevant prescriptive requirement(s). The performance criteria shall provide a level of safety not inferior to the relevant prescriptive requirements contained and performance criteria shall be quantifiable and measurable;
- Detailed description of the alternative design and arrangements, including a list of the assumptions used in the design and any proposed operational restrictions or conditions;
- Technical justification demonstrating that the alternative design and arrangements meet the safety performance criteria; and
- Risk assessment based on identification of the potential faults and hazards associated with the proposal.

11.5 The Red Ensign Group Yacht Code chapter 1.10 covers exemptions stating that exemptions shall be granted only by the administration and applications shall be supported by justification for the exemption. The granting of exemptions shall be limited by the extent to which the international conventions allow and shall in all instances be regarded as exceptional.

11.6 The Red Ensign Group Yacht Code chapter 1.8 covers Equivalent Arrangements which also need to be submitted to the Administration for approval. Deviations from the requirements of the Code may be considered by the Administration, provided that the equivalence:

- is based on the underlying Convention requirements;
- incorporates increased requirements to balance deficiencies and thereby provide an equivalent level of safety to the chapter concerned;
- meets the intent of the requirements concerned and the objectives of the chapter;
- incorporates any specific requirements in the chapter's 'Additional Equivalence Considerations' section; and
- where necessary has successfully undergone testing to the satisfaction of the Administration; or

has successfully undergone, an engineering analysis, evaluation and approval by the Administration.

12. NAVIGATION

12.1 The SOLAS chapter (V) has a widespread application to almost all ships on all. Administrations can determine to what extent the provisions of regulations 15, 16, 17, 18, 19, 20, 21, 22, 23, 24, 25, 26, 27 and 28 do not apply to ships below 150 gross tonnage engaged on any voyage, ships below 500 gross tonnage not engaged on international voyages and fishing vessels. General exemptions from the requirements of regulations 15, 17, 18, 19 (except 19.2.1.7), 20, 22, 24, 25, 26, 27 and 28 can be applied to ships without mechanical means of propulsion.

12.2 Individual ships may be granted exemptions or equivalents of a partial or conditional nature, when any such ship is engaged on a voyage where the maximum distance of the ship from the shore, the length and nature of the voyage, the absence of general navigational hazards, and other conditions affecting safety are such as to render the full application of the chapter unreasonable or unnecessary, provided that the Administration has taken into account the effect such exemptions and equivalents may have upon the safety of all other ships.

12.3 Navigational equipment has already benefited from many technological developments especially e-navigation with a view to improving the efficiency of the whole maritime shipping operation as well as reducing the associated risks by enhancing navigational systems and communications. The development of integrated bridge systems has been an area where better integration of human and machine interfaces has occurred, however there are still few fully integrated systems in existence which can still cause frustration to the users and associated risks. Ergonomics and also the dangers associated with information overload have not always been properly addressed. All too often alarm systems can be too complex and occasionally lead to operators missing serious alerts.

12.4 The carriage requirements for such items as; Automatic radar Plotting Aids (ARPA), Automatic Identification Systems (AIS), Bridge Navigational Watch Alarm Systems (BNWAS), Long Range Identification and Tracking (LRIT), Electronic Chart Display and Information System (ECDIS) have all presented challenges at implementation stage with stringent specifications being necessary to ensure compatibility and safe operation.

12.5 Aids to Navigation, ships routing and weather routing services are amongst the areas where technology developments have greatly aided the promulgation and presentation of information to mariners and allow for optimization of vessel operations to improve the overall efficiency of the transportation operation. Search and rescue distress signaling etc. are areas where improvements have been made to help if things do go wrong.

12.6 Information is greatly enhanced by the improved communication systems now available which also enable much improved monitoring of vessel location and performance by ship operators and also flag states.

12.7 Hydrographic activities are already an area where autonomous vessels as well as remotely operated ones are being used to an increasing extent and as these develop from relatively small boats so the need to further accommodate these developments within regulation will increase.

12.8 The "B0"–conditionally and periodically unmanned bridge possibilities could well eventually develop to enable the removal or moving navigational functions away from the traditional location on board which will give potential reduction in superstructure heights and aerodynamic improvements or on passenger carrying vessels enable greater development of panoramic lounges etc.

12.9 Sound reception systems are already accepted on enclosed bridges. CCTV already permitted on some vessels for covering blind spots and assisting berthing as have berthing radars. Whilst care is needed in ensuring that human interpretation of information presented on screens does not deceive the eyes there is great potential for more optimal positioning of optical sensors leading to reduced lookout needs which could cut manning or improve service. The use of reliable optical sensors and CCTV could indeed lead to improvements in visibility for the bridge staff and obvious benefits to safety through better lookout. The ever-increasing reality of ship simulators is a good indication of what can be achieved.

12.10 There should be at least two independent means for all measurement and situational awareness technologies. There must be proper monitoring and diagnostics of all navigation sensors and equipment. Such requirements should lead to the navigation of ships being conducted in a manner at least as safe as currently exists. When introducing technology there is obviously a need to ensure that safety is not compromised, however it is also necessary not to introduce too onerous requirements that are in excess of those currently expected of humans performing the function.

13. CONCLUSION

There is plenty of opportunity for development of autonomous ship solutions within the constraints of the current conventions all of which will help promote the shipping industry as a vital part of International Trade which is safe, economically viable and environmentally friendly

14. AUTHORS BIOGRAPHY

Christopher Balls, a Master Mariner who spent 25 years at sea before coming ashore as a Marine Surveyor with the UK Maritime and Coastguard Agency and then spent four years as Chief Marine Surveyor for the Antigua and Barbuda Department of Marine Services and Merchant Shipping before moving on to Cayman Maritime as Principal Surveyor in the George Town Headquarters , with responsibilities including ensuring the regulatory compliance of the country's fleet and visiting vessels.

15. AKNOWLEDGEMENTS
The aforementioned IMO and ILO convention and related texts.

DEVELOPING THE MARLAB MARINE AUTONOMOUS SURFACE SHIPS DATA PROJECT

R. J. H. Blazejczyk, Solis Marine Consultants, UK
W. J. Ellison, MariTrace, UK
S. Hindley, Solis Marine Engineering, UK

SUMMARY

Central to the UK Maritime Coastguard Authority (MCA) MARLab project is the data workstream. The purpose of this study is to define the parameters for a single, common, interactive and scalable data hub to share mission critical navigation data, facilitating the safe testing of smart and autonomous vessels of less than 24 metres in length in UK waters. The aim of this data sharing facility is to unlock the potential for the future development of MASS (Marine Autonomous Surface Ships). In the process, MARLab maps the requirements for a legal and regulatory framework and addresses significant barriers and concerns which principal players say are hindering sector development and growth.

1. INTRODUCTION

According to industry estimates, there are believed to be more than 1,000 autonomous vessels operating in international waters and around 53 or so organisations contributing to various regulatory groups working to help MASS safely co-exist with manned shipping.

In the past six years, remote and unmanned vessels have successfully navigated tens of thousands of incident free operative days. Next generation MASS is no longer a purely academic exercise.

Principal MASS players such as BAE Systems, Thales, L3-Harris and Ocean Infinity/Armada are now looking to develop at speed. Key drivers include general survey services, pipeline / cable survey work, oil and gas, and the offshore wind industry which is set to grow three times its current capacity by 2030[1]. MASS is identified as key to keeping down the costs of energy supply through autonomous maintenance solutions.

Research bodies have identified a wide range of autonomy at differing technology readiness levels (TRLs). Most MASS systems have options for remote piloting and some situational awareness capabilities. However, designers, builders and operators only have industry or certifying authority best practice to utilise as a reference point at this time.

Now customers would like to move in the direction of true autonomy, whereby USVs (Unmanned Surface Vessels) can operate without human oversight. Currently, an operator with full situational oversight is considered a requirement for safe operation. Customers expect a high maturity of MASS, supported by significant evidence to demonstrate safe operation.

The availability of good quality data, however, is perceived to be inaccessible when in fact stakeholders claim it should be made more freely available by the Government. The current scenario, in which operators have to rely on several different data sources, is posing a barrier to the development of the MASS industry compared to autonomous land and air propositions.

Last year, a more consultative approach was taken to understand the barriers and unlock the necessary data required to progress towards Maritime 2050. This is the Department for Transport (DfT) strategy, a bold vision of what the Government hopes the UK's maritime sector will look like 30 or so years from now. This strategy sets out commitments to ship automation and remote operation.

To advance MASS, it was evident that we needed to create a comprehensive visualisation of remote and local data upon which decisions could be made. We anticipated the need for a free to access Maritime Data Hub as a repository for multiple sources and data types, made available to the smart and autonomous shipping industry for research, planning and operational use.

2. THE CASE FOR MASS

The case for MASS is clear. MASS have the ability to remove humans from dangerous, dirty and mundane activities such as hydrographic surveys in coastal zones. MASS are often smaller than conventional manned systems so potentially offer the ability to do things more cheaply and more flexibly. They also offer the potential to do things that conventional systems can't do, for example, very shallow water bathymetry.

Because they can be smaller in size and often use alternative propulsion systems, MASS are less costly to run and have better environmental credentials than conventional systems. The remote nature of MASS also means that they can be deployed for longer periods of time with minimal manning requirements and stay in residency in a particular location if required.

Highly capable simulation environments have built up a body of evidence, providing the necessary assurances to regulatory bodies. Extensive realworld testing is underway at several UK locations including Langstone Harbour, Portsmouth, Portland Harbour and Plymouth Sound, and there are exploratory talks involving academic institutes such as Plymouth, Southampton and Exeter Universities.

Testing facilities include:

- Electrical and mechanical labs
- Dry dock simulated environments
- Shallow water testing environments
- Deeper water environments
- SubSea structure offshore infrastructure

The Offshore Renewable Energy Catapult, (OREC), has a working relationship with Vattenfall's Aberdeen Bay (European Offshore Wind Deployment Centre EOWDC) specifically for demonstration purposes of new technologies.

Smaller remotely controlled vessels are deliberately compact and light and are designed to minimise the risk to other marine users. It is recognised that offshore wind tends to take more of a chance on less extensively tested technologies that would be expected from other industries. The assets are typically unmanned, have no explosive qualities and therefore represent a low risk of human loss of life from a potential MASS incident.

OREC is seeing an increase in residency MASS mothership systems incorporating elements that enable operational and maintenance tasks. And the MIMRee system, set up to develop the world's first fully autonomous offshore wind shore repair system, deploys Thales' Halcyon autonomous vessel along with robotics and a drone system.

With regards to risk assessment and the mitigation of risk, The University of Plymouth has received substantial funding from Research England to develop a Maritime Cyber Security Lab that will effectively see the University become the National Centre of Excellence for Maritime Cyber.

The University says it would welcome engagement with the MARLab project to apply their leading thinking around Risk Assessment and Mitigation of Risk in this area.

Despite multiple advances however, operators say that a lack of international recognition, internationally agreed standards for semi-autonomous or fully autonomous vessels, and free to access data, limits their ability to safely interact with manned vessels.

3. DATA PROTOTYPE DEVELOPMENT

Working with Solis Marine, MariTrace has developed a system to demonstrate how multiple data sources from a single port can be consolidated into a single location, making them available for Marine Autonomous Surface Ships (MASS) to access the data on demand.

The system has been developed as a Minimum Viable Product (MVP) which is intended to:

- demonstrate the concept of a data hub for this purpose;
- give MASS developers access to this data hub and the datasets available through it, allowing us to gather valuable feedback;
- explore system architecture and technology solutions for full scale development; and,
- identify potential design considerations or sticking points for full scale development.

The MCA in conjunction with Solis Marine chose Portland Harbour as the ideal test area because it is big enough for a MASS test area but not so large that datasets for this viability study become unwieldly.

Portland has many different data sources to manage its port on a day to day basis. These have been harmonised into a single portal alongside other data which has been supplied by the MCA and other bodies.

3.1 DATA HUB SYSTEM ARCHITECTURE

The system ingests multiple data sources and converts the raw data into a standardised structure to allow for fast storage and retrieval. The data is retrieved by the end user through a series of Application Programming Interfaces (APIs).

-The system includes a dashboard to monitor the activity of each of the data sources and APIs, and a suite of help files providing technical details on how to integrate with the APIs.

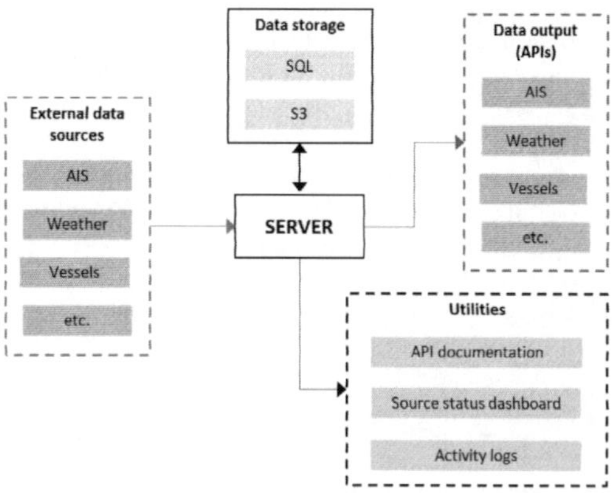

Figure 1: MVP Architecture

The key components of the system are:

3.1 (a) External data sources

The process for ingesting each source depends on the architecture of the system it originates from.

3.1 (b) Server

The server runs the core functionality that handles all processes throughout the system, including:

- Converting the external data sources into a standard format for easy storage and retrieval
- Archiving and retrieving the data as required
- Responding to end user requests for data
- Hosting the dashboard and documentation pages
- Writing input and output activity logs

3.1 (c) Data storage

All external data received by the system is archived in a fast access database.

3.1 (d) Data output

The data is accessed by the end user via an Application Programming Interface (API) specific to each data source.

An API is a structured format for connecting two systems, where both the requester and responder know the exact format of the data being sent and received. This allows programmers to make the data available to any other programme on demand.

3.1 (e) Utilities

There are three main components to the utilities:

- API documentation, giving any programmer the reference materials needed to integrate with the APIs
- The source status dashboard provides a visual indication of how each data source is performing, including details of the latest data received from them
- Logs of all API requests made by end users to gauge usage

3.2 TECHNOLOGIES USED

The technology stack was determined based on MariTrace's experience handling, sorting and processing large quantities of maritime data, and consultation with developers of other similar platforms including the team behind the Data & Analytics Facility for National Infrastructure (DAFNI) project.

The key technologies used are as follows:

3.2 (a) Server

MS SQL Server. Cloud based. Physically located UK South.

3.2 (b) Storage

Currently 250GB although this can be increased.

3.2 (c) Database architecture

SQL Tables-based.

3.2 (d) API language

C# with various libraries.

3.2 (e) Web hosting

MS Azure cloud. Physically located UK South. App services.

3.2 (f) Dashboard and interface

Written with C# back-end with various libraries. JavaScript front-end on aspx .NET platform with Telerik controls for graph visualisation and MapBox JS API.

3.3 DATA SOURCE DASHBOARD

The dashboard system provides a visual understanding of the status of each data source. The dashboard is accessible to anyone with a valid username and password, for the purposes of the MVP the user access details are maintained by MariTrace.

It should be noted that the dashboard access credentials are separate from the API authentication credentials.

The dashboard is hosted at the following URL: https://www.marlab.co.uk/

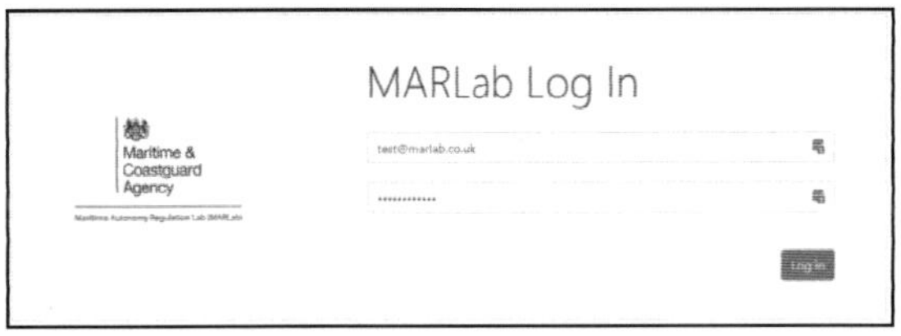

Figure 2: MARLab log in page

After logging in, the user arrives at the 'Data Source Health' page, which lists all the available data sources, and provides details on the update frequency and latest updates from each one. It includes a traffic light status for each data source, indicating:

- Green – the data source is updating as expected
- Amber – the data source is experiencing delays, but not necessarily of major concern
- Red – the data source has not updated for longer than expected

It should be noted that there is an API available which returns the same information displayed on the dashboard, allowing MASS systems to automatically check whether a data set is reliable in real time.

Full details can be found at the following URL: https://api.marlab.co.uk/Help/Api/POST-api-DataSetStatus

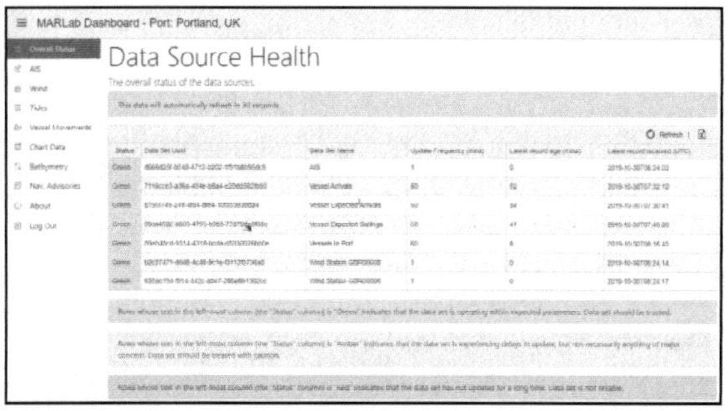

Figure 3: Data Source Health

As well as the overall status, there is a separate page for each data source providing specific detail relevant to that source. The design of each page depends on the characteristics of that data set, but would typically include:

- Details about the location of the source
- The historical and latest data from that source
- An indication of the number of data requests made to that source, and the volume of data being returned in each case.

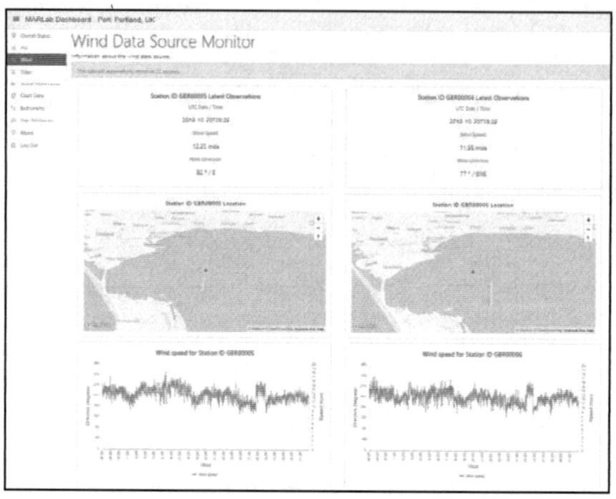

Figure 4: Wind Data Source Monitor

3.4 API ACCESS AND HELP FILES

The data is delivered to end users via API. The API is accessed through a URL that allows users to submit a request, which then responds with the required data.

Access to the API is authenticated and monitored by means of a username and password that is submitted to the API with each request. For the purposes of the MVP the credentials are managed by MariTrace.

Documentation outlining the structure and technical details for each API has been provided as an online resource, which is standard practice in the software development community.

The API access point and full documentation is hosted at the following URL: https://api.marlab.co.uk/

The API site documentation page lists all APIs currently available:

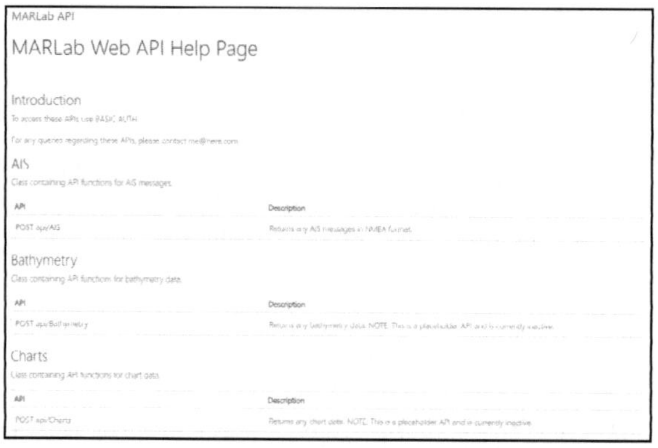

Figure 5: MARLab web API help page

Each API has a dedicated page showing all details required by the end user, in order to submit a request. This includes example code which the developer can copy. The API accepts two API languages, JSON and XML, and examples are given in both:

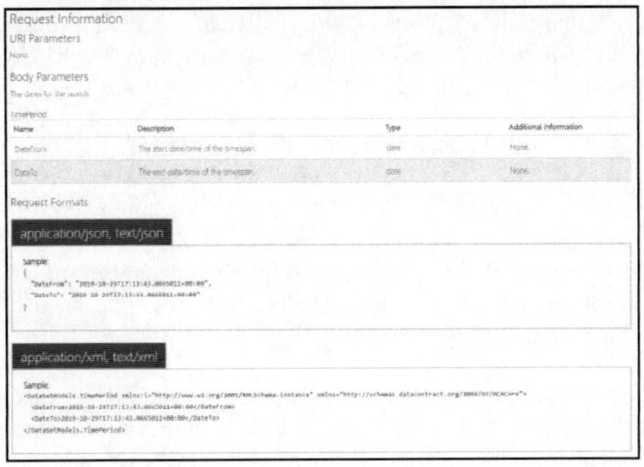

Figure 6: Example API page

Similarly, the documentation describes the response from the API, including sample responses in both JSON and XML:

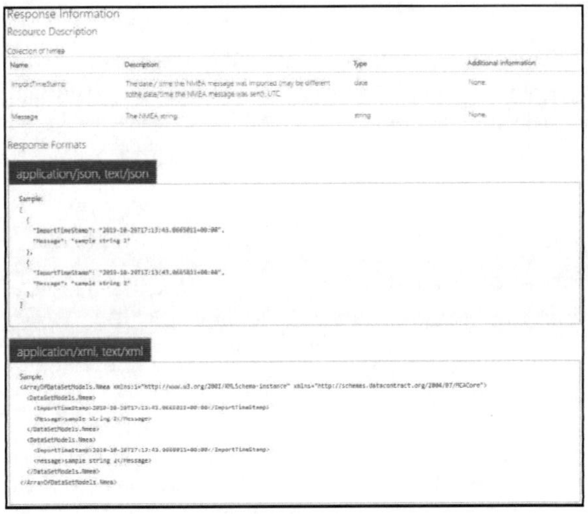

Figure 7: Sample API Responses

3.5 DATASETS

For the purposes of the MVP, a significant number of data sets for Portland Harbour have been made available. Full technical details for each of the data sets is available at the API help page (https://api.marlab.co.uk/Help).
The 'API Request' indicates the details of the query the user submits to the API, and the 'API Response' outlines the results that are returned.

It's important to reiterate that stakeholders invited to test the MVP require the software capability to write the necessary code to access the API although template scripts can be provided. User testing continues until the end of March 2020 as we add more data to the prototype hub.

The MVP now has the following data sources available:

3.4 (a) Automatic Identification System (AIS)

Live and historical navigation data transmitted by all vessels in or around the port of Portland. Historical data is provided by the MCA. Live updates are provided by Pocket Mariner Ltd.

3.4 (b) Vessel Movements

The latest information from the Portland Port Authority vessel scheduling software. This data includes four different sets:

- Vessel arrivals
- Vessel expected arrivals
- Vessel expected sailings
- Vessels in port

A separate API has been set up for each of the four subsets.

3.4 (c) Navigation Advisory

Announcements from the port authority relating to warnings, alerts or restricted areas. Data structure provides flexibility to allow for a broad range of announcements to be made.

3.4 (d) Wind

Live and historical wind speed and direction, taken live from Anemometers located in Portland Port (taken from https://weatherfile.com/)

For the MVP this is available from two observation stations.

3.4 (e) Tide

Tide height forecast for Portland, supplied by the UKHO.

3.4 (f) Weather Forecast

Latest weather forecast for Portland, as issued by the UK Met Office.

3.4 (g) GeoPackage

A single file containing the latest navigation data from the UKHO for Portland. Contents are in the following categories:

- Natural Sea Floor
- Navigation Marks
- Artificial Sea Floor
- Topography
- Tides
- Bathymetric
- Marine Services Signalling Stations
- Port Harbours
- Surface Installations
- Regulatory Limits
- Geographical Area
- Inland Waterway
- Marine Aquaculture
- Meta Features
- Coast
- Routing Measures
- Magnetics
- Wrecks Areas
- Wrecks Points
- Restrictions
- Marine Management Areas

4. GAP ANALYSIS

A Gap Analysis is being completed in order to understand what government held data could be made available and what the MCA and industry aspirations are for the use of that data so as to define the resources available to make the connection between government and the industry in this respect. This part of the project is being conducted in two stages with the first describing data sets held by the MCA and Portland Harbour Authority that have been made available for the MVP. This section of the study also reviews existing marine data sharing platforms.

The second part of the Gap Analysis study is ongoing at the time of writing and is designed to address in greater detail the data available from other government organisations including the UKHO and the Met Office.

Data from the UKHO is now available through the MVP as the geo package source. Met Office meteorological 'spot' data for a location in Portland is also now available through the MVP.

Further work in this area has included discussions with the Geo Data department at the University of Southampton leading to an opportunity to potentially include data from the Channel Coast Observatory portal. This data source is seen as potentially highly valuable to the MARLab data offering and complements the existing data sets already available.

A final report for this element of the project will be delivered at the end of March 2020 with a view to ensuring that the gap analysis helps to form a future work scope for the continuation of the MARLab portal.

The second phase of the consultation work is ongoing and consists of individual feedback sessions with those stakeholders that are currently testing the data portal and a stakeholder workshop held in parallel with the Policy Lab regulatory workshop.

The purpose of this workshop will be to ascertain how and where the MASS industry wants to see the MARLab data work develop beyond the scope of this project.

5. STAKEHOLDER ENGAGEMENT

A broad mix of significant stakeholders including MASS operators, developers, research bodies and port authorities have participated in the viability study, contributing to the Gap Analysis and prototype testing exercises.

As expected, there has been some frank feedback regards the leadership shown to date by the International Maritime Organisation (IMO) and other bodies.

Stakeholders have also pointed to a lack of customer education and user communications to build necessary trust and industry confidence. At its most basic, they identified a need for a system which verifies that a MASS vessel is capable of complying with collision regulations autonomously.

Communications between MASS operators and on the water users to prevent unexpected actions and avoidable mishaps if and when on the water users encounter USVs was raised by several concerned stakeholders. To this end, stakeholders are looking to the MCA to develop a demonstration of best practice as guidance to new operators.

Low cost, high bandwidth satellite communications were also pinpointed as bringing the most benefit to MASS operations; more reliable, robust greater bandwidth communications infrastructure to increase range to cover the whole globe.

Complex safety related behaviours and critical variables, such as weather and tidal systems, have naturally been highlighted as concerns, along with the time and cost of at sea trials at scale to grow confidence.

Others cited the development costs involved, the availability of representative assets, the reliability and the resilience of vessels and the costs of retrofitting them as barriers to testing and development.

With regards to the regulatory environment and current absence of a firm regulatory framework, stakeholders are seeking clarity on the type and level of evidence required by regulators to demonstrate assurance of safety and global alignment on regulations and standards. An amendment to The Workboat Code Edition 2 Amendment 1 is underway.

5.1 REQUIREMENTS

The viability study establishes a need for the provision of freely available, quality data from assured sources. Available data is regarded as inaccessible and expensive, posing a barrier to the speed of development of MASS.

Critical requirements included situational awareness data i.e. the use of AIS to increase understanding of other vessels. Stakeholders also pinpointed the need for highly reliable obstacle data, vessel insurance information and 100% reliable tracking and identification of live obstacles.

The top four datasets which were prioritised by stakeholders as being the most important are all included on the Portland Harbour prototype. These are AIS data; tidal data; UKHO geo spatial data and a hydrographic model of UK waters.

MASS stakeholders require an effective regulatory framework for accrediting vessels for wider operation. Some highlighted the lack of clarity with regards to the regulatory environment and the absence of a firm regulatory framework. They require greater transparency in relation to the type and level of evidence required by regulators to demonstrate assurance of safety; and are seeking global alignment on regulations and standards.

Stakeholders are expecting the MCA and Harbour Authorities to take as supportive an approach as is practicable. Lack of leadership from the International Maritime Organisation (IMO) and other bodies was a concern for a number of stakeholders. It was suggested that MCA should develop a demonstration of best practice as guidance to new operators.

Other barriers to testing were identified, including a lack of government funding, development costs and the availability of representative assets. The time and costs of at sea tests was raised along with concerns with regard to the reliability and the resilience of vessels and the cost of retrofitting them.

There was a call for reduced sea trial costs and additional financial subsidies/incentives for traditional companies to invest more in MASS technology. Stakeholders want to see support for enterprise, in particular that which seeks to provide low cost satellite data communications to marine users because a lack of low cost, high bandwidth satellite communications is seen as a major barrier to testing MASS.

Stakeholders require a clearer understanding of the pathway to MASS and our consultation establishes a fundamental need for better communications infrastructures. Respondents identified customer education as another key requirement and said that communications with potential users was necessary to build public confidence. An overwhelming majority of respondents (16 yes versus two maybe) are in favour of a workshop to explore how the MCA and the DfT can support industry data needs for MASS operation. As such, a workshop is planned towards the concluding part of the project.

Beyond data, greater engagement with UK universities to enable the sharing of international practice was proposed, for example, the University of Plymouth has received funding from Research England to develop a Maritime Cyber Security Lab and says it would welcome engagement with MARLab to share its leadership in the field of mitigating risk in relation to MASS.

6. CONCLUSIONS

The MARLab data stream study has identified a range of requirements to address costs, performance, growth and protocols to make UK waters safe for MASS testing.

Feedback from the data led MVP shows that stakeholders would like to widen the scope of the prototype to include waters and ports outside of the current Portland digital test area.

Work is now underway with partners to progress towards the development of a wider, intelligent database that captures every aspect of the maritime industry which includes data on location, weather, emissions, fuel and journey times and routes. Such a resource will help the government meet the objectives set out in Maritime 2050.

6.1 WIDENING ACCESS TO DATA

From our work to date we know that improved cost and performance of satellite broadband systems, together with integrated and centralised access to multiple data sources including fishing VMS, AIS, weather etc, stands to benefit not just the MASS community, but other stakeholders in the marine sector as well.

To widen access to the data available one priority would be to develop the graphical user interface (GUI) to allow access via an online form and a direct download as an alternative to using APIs. Software developers recognise that APIs are now the standard method for accessing large data sets, but personnel without the coding expertise required to work with APIs require a more mainstream approach.

Developing a more sophisticated GUI is one of the more straightforward areas for any future expansion of the MARLab portal.

6.2 DEVELOPING NEW INTERFACES

Any communications protocols and technologies which allow MASS to communicate with each other, as well as land based centres, should ideally be developed in a way which is accepted and adopted by the global industry.

Looking to a future world of smart ports and the transfer of goods autonomously between marine to road and rail, some interface with land-based systems is required.

This is already an aspiration in Plymouth, and so it would seem sensible to discuss this now to avoid any duplication of effort in the future.

We would also look to work with other stakeholders to encourage collaboration and partnership to pioneer the coherence between surface, sub surface, air and land autonomous systems.

Indeed, there is a general expectation for improved collaboration and partnerships to pioneer the coherence between surface, sub surface, air and land autonomous systems.

Lessons can be learnt from other sectors such as aerospace and road transport and there is an expectation from stakeholders that the UK marine sector should be taking the lead.

9. REFERENCES

1. DEPARTMENT FOR BUSINESS, ENERGY AND INDUSTRIAL STRATEGY, *Gov.UK* March 6, 2019

10. AUTHORS BIOGRAPHY

Rosalind Blazejczyk holds the current position of Managing Partner at Solis Marine Consultants, is a naval architect, chartered engineer and a member of the Panel of Special Casualty Representatives (SCR) at Lloyds. In addition to her role with the MARLab project, she works as a consultant in relation to shipping incident investigation, salvage and wreck removal. Her previous experience covers a number of complex and politically sensitive salvage and wreck removal operations worldwide.

William Ellison is a naval architect and chartered engineer who specialises in applying maritime data to a wide range of applications. As a consultant he has provided expert advice to the Supreme Court of Singapore on the use of ship tracking data in collision litigation and working with MariTrace has helped develop data services to understand and analyse the flow of shipping cargo around the globe.

Simon Hindley holds the current position of Managing Director at Solis Marine Engineering Ltd. He is a naval architect responsible for engineering design and analysis services. His previous experience includes work for the UK MOD and a range of subsea marine renewable energy projects from concept design to offshore deployment and maintenance.

AUTONOMOUS SHIP: MANOEUVRING AND CONTROL IN LABORATORY ENVIRONMENT

A. Chandra Dubey, V. A. Subramanian and P. Krishnankutty, Indian Institute of Technology Madras, India

SUMMARY

The development of the autonomous ship needs prior knowledge of the hydrodynamic motion behaviour and stability in external wave conditions. Such studies can be classified under the two main areas of i) Sea keeping ii) and Maneuvering. Both studies involve knowledge of stability, motion and control. Knowledge of ship steering dynamics is required to evaluate the manoeuvrability and for the design of the autopilot and navigation system. This paper presents the study of the control, navigation and communication protocol in the laboratory environment by simulating different ship manoeuvers in several wave conditions. The developed autonomous on-board system equipped with main computer, suitable electronics, sensors, data acquisition system and Wi-Fi based communication system. The Wi-Fi based communication between shore station and the ship model facilitates the user to access the data over any web browser making it independent of operating system and enables to do Internet of Things (IOT). Various steering control designs such as PID and LQR steering controllers are simulated and compared on a mathematical model obtained from standard manoeuvring tests. A preliminary physical simulation in the laboratory environment is performed to verify the communication protocol and the model response in different wave conditions.

NOMENCLATURE

Symbol	Definition (Unit)
m	Mass of the ship model (kg)
u	Speed of the model in x direction (m/s)
v	Speed of the model in y direction (m/s)
I_z	Moment of Inertia about z axis (kg m^2)
X	Force in x direction (N)
N	Yaw moment (N-m)

1. INTRODUCTION

Currently, the direction of global research is in the area of developing autonomous ship such as cargo ships, container ships, ferries etc. [1]. The autonomous navigation of the vessel helps to improve sailing efficiency and economize on running cost by saving manpower cost. Recent advances in communication protocol, controllers, robust trajectory tracking control, high precision sensors, in artificial intelligence and the Internet over things (IOT), the autonomous ships are going to be reality in the near future. The best example of such implementation is the Rolls-Royce Company effort in designing and implementing container ships in the real world.

Recent studies show the potential of increasing autonomy in shipping field, as compared to the autonomy on land and air, the autonomy in shipping is fledgling. Both the European research project MUNIN, [2], the zero-emission, unmanned, short sea concept of DNV [3], and a cooperation between among others Rolls-Royce and DNV GL in the AAWA project, Rolls-Royce [4], are conducting research into the topic of autonomous shipping. The focus here lies on longer distance autonomous and unmanned sailing outside ports, presenting impressive figures on cost savings due to the absence of crew on board [5]. On the other hand there is a noticeable drive towards increasing autonomy in the optimization of inter-terminal transport of containers by using autonomously guided vehicles (AGVs) within the port of Rotterdam to reduce costs and waiting times [6,7]. Here the focus was on autonomous transport over land, with in one case also on waterborne AGVs, [8].

Studies on maneuverability and steering control and design of autopilots require physical simulations in laboratory environment. Even with the availability of computational tools for simulations, it is necessary to validate the results through physical models. The standards on maneuvering requirements of vessels set by regulatory bodies such as the International Maritime Organization (IMO) also make it necessary to validate the maneuvering capability through simulations and physical experiments on scaled down models. The ship dynamics in 6-DOF is determined from the first principle using Newtonian Mechanics [9]. The assessments based on nonlinear and coupled hydrodynamics coefficients involved in the equations of motion are fairly complex to obtain through experimental or numerical techniques[10-13]. The 6-DOF motion is decoupled into two sets namely, the vertical plane motions which include roll-pitch and heave sub-system and horizontal motions which include the surge-sway-yaw sub-system. The latter helps in the study of maneuvering of ship and surface vehicles. Nomoto presented a simplified steering model of a ship for designing autopilots based on input-output based transfer function [14]. The coefficients of Nomoto transfer-function is obtained using self-propelled scaled down ship models by performing free running tests and standard maneuvers. Based on the Nomoto transfer function, different steering control models are evaluated for designing the autopilot system.

2. SYSTEM DYNAMICS AND CONTROL DESIGN

2.1 SURGE MODEL

The surge sub-system essentially requires correlation of propeller rate of revolution to the ship speed. Estimates can be made either analytically or more reliably by careful tests on scaled down model in a towing tank. In all cases the final validation is required through physical model tests. The model is prepared with prime mover BLDC motor and controls and connected to the propeller through shafting. The tests are fairly straight-forward and consist of revving the propeller to different steady rotational speeds and noting the steady speed achieved in each case. Figure. 1 shows the obtained results. A polynomial curve gives the relationship between propeller rpm and speed.

2.2 YAW SUB-SYSTEM

The coupled nonlinear differential equations of motion in 6-DOF are derived using the Newton-Euler equations. Using the horizontal plane motion equations and neglecting the vertical plane motion equations1 give:

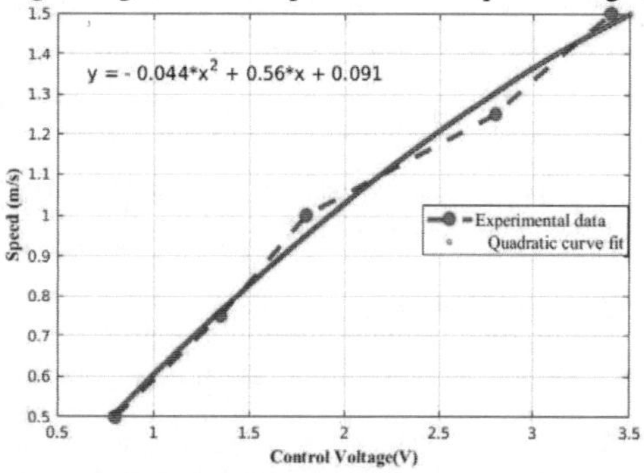

Figure 1: Speed calibration for the candidate hull

$$\begin{aligned} &m[\dot u - vr - x_g r^2 - y_g \dot r] = X \\ &m[\dot v - ur - y_g r^2 - x_g \dot r] = Y \\ &I_z \dot r + m[x_g(\dot v + ur) - y_g(\dot u - vr)] = N \end{aligned} \tag{1}$$

where m is the mass of the ship model, u is the surge velocity in the x-direction, v is the sway velocity in the y-direction, r is the yaw rate, I_z is the moment of inertia about the z-axis, x_g and y_g location of the centre of gravity, X indicates surge force, Y indicates sway force, and N represents the yaw moment. Under assumption of constant model velocity, the above equation further reduces to the yaw motion equations given by:

$$\begin{aligned} &m[\dot v - u_0 r - y_g r^2 - x_g \dot r] = Y \\ &I_z \dot r + m[x_g(\dot v + ur) - y_g(\dot u_0 - vr)] = N \end{aligned} \tag{2}$$

Sway and yaw are dependent functions as below:

$$\begin{aligned} Y &= Y(v, r, \delta, \dot v, \dot r) \\ N &= N(v, r, \delta, \dot v, \dot r) \end{aligned} \tag{3}$$

where δ is the rudder deflection.

The linearized maneuvering equations are:

$$\begin{bmatrix} m - Y_{\dot v} & mx_g - Y_{\dot r} \\ mx_g - N_{\dot v} & I_z - N_{\dot r} \end{bmatrix} \begin{bmatrix} \dot v \\ \dot r \end{bmatrix} = \begin{bmatrix} Y_v & Y_r - m \\ N_v & N_r - mx_g \end{bmatrix} \begin{bmatrix} v \\ r \end{bmatrix} + \begin{bmatrix} Y_\delta \\ N_\delta \end{bmatrix} \tag{4}$$

where $Y_{\dot v}$ and $N_{\dot r}$ are the added mass and added mass moment of inertia respectively. $N_{\dot v}$, $Y_{\dot r}$, Y_v, Y_r, N_v, N_r and Y_δ, N_δ are the damping coefficients and hydrodynamic derivatives related to rudder, respectively.

3.3 INPUT-OUTPUT BASED FIRST ORDER NOMOTO MODEL:

The first order Nomoto model is derived from the linearized maneuvering model. Under Laplace transformation it yields:

$$\frac{r}{\delta}(s) = \frac{K(1 + T_3 s)}{(1 + T_1 s)(1 + T_2 s)} \tag{5}$$

where T_1 and T_2 are the indices of stability on course and quick response in steering respectively, T_3 represents the contribution of steering speed in initiating a turning motion and K is an index of turning ability . The effective time constant is defined as, $T = T_1 + T_2 + T_3$ where T is obtained from certain definitive maneuvers such as the turning circle or zigzag tests. The steering motions of the ship can be approximated by the first order Nomoto model given by:

$$T\ddot\psi + \dot\psi = K\delta \tag{6}$$

The Laplace transformation of the above equation can be represented as:

$$\frac{r}{\delta}(s) = \frac{K}{(1 + Ts)} \tag{7}$$

Where $r = \dot\psi$ which gives,

$$\frac{\psi}{\delta}(s) = \frac{K}{s(1 + Ts)} \tag{8}$$

For this example, the values of K and T were obtained by conducting the standard zigzag test on a scaled down free running model of a candidate coastal research vessel in a wave basin facility. The values obtained are K = -0.212 and T= 1.785. The turning circle test and zigzag tests are simulated using obtained transfer function as shown in Figure 2 and Figure 3 respectively.

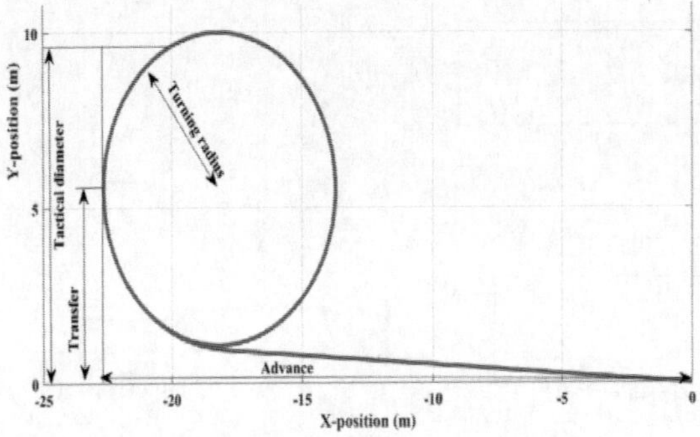

Figure 2: Turning circle test simulated using first order transfer function obtained from manoeuvring tests

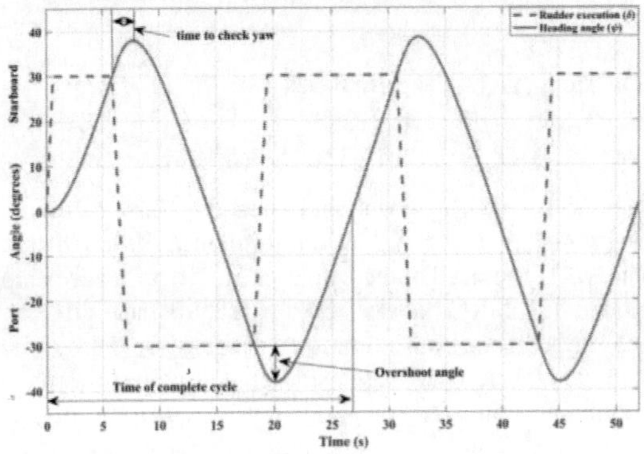

Figure 3: Zigzag test simulated using first order transfer function obtained from manoeuvring tests

3. STEERING CONTROL DESIGN

The control objective is to ensure that the vessel keeps desired course despite external disturbances acting on it. The steering control should drive the error e to zero i.e. $\lim_{t\to\infty}(\psi_d - \psi) = 0$, where ψ and ψ_d are the current heading angle and the set desired heading angle, respectively. In the scaled down laboratory tests these angles are measured on an on-board measurement device namely the motion reference unit (MRU). The control logic of the two control options namely the PD controller and the LQR controller are as follows:

3.1 PD CONTROLLER

A PID based steering control is very popular and often adopted in the ship maneuvering, due to their functional simplicity and robust performance in a wide range of operating conditions. To design the Proportional Derivative (PD) controller in the frequency domain, the control law is given by:

$$\delta_r = K_p(\psi_d - \psi) - K_d \times r \tag{9}$$

where, δ_r is the actuator signal to control the rudder, K_p is the proportional gain, K_d is the differential gain, ψ_d is the desired heading, ψ is the current heading and r is the yaw rate.

The transfer function of the PD controller in frequency domain is represented as:

$$G_c(s) = K_p + K_d s \tag{10}$$

The mathematical simulation is performed to tune the PD controller in Simulink as shown in Figure.

3.2 LQR CONTROLLER

The Linear Quadratic Regulator (LQR) is another optimal control technique which is a full state feedback controller. The LQR controller uses the state space based approach. By conducting the turning circle test with a convenient rudder angle and measuring the steady yaw rate, one obtains the first order model transformed into state-space as given in (9). The feedback gain is a matrix K_l, implemented as $u = -K_l(x - x_d)$. The closed loop dynamics is then written as:

$$\dot{x} = (A - BK_l)x + BK_l x_d \tag{11}$$

x_d represents the desired vector contains heading and heading rate and serves as an external input to the closed-loop system and $A = \begin{bmatrix} 0 & 1 \\ 0 & -\frac{1}{T} \end{bmatrix}, B = \begin{bmatrix} 0 \\ -\frac{K}{T} \end{bmatrix}$

The optimal LQR problem is defined more generally and consists of finding the control input that minimizes:

$$J_{LQR} = \int_0^\infty x(t)'\bar{Q}x(t) + \rho u(t)\bar{R}u(t)dt \tag{12}$$

where Q and R are weighing matrices. The value of K_l is calculated based on the value of Q and R matrices to match the desired response using inbuilt function *lqr(A, B, Q, R)* coded in MATLAB. The results based on the two controllers are compared in Figure 4. Figure 5 explains the Simulink block diagram for PID and LQR control technique

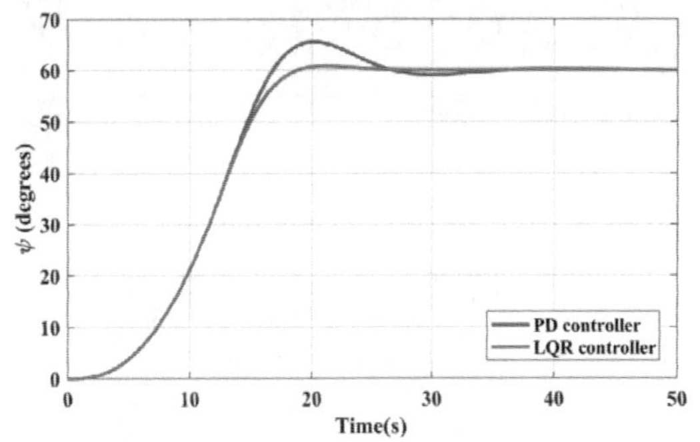

Figure 4: Comparison of heading response of PID and LQR controller

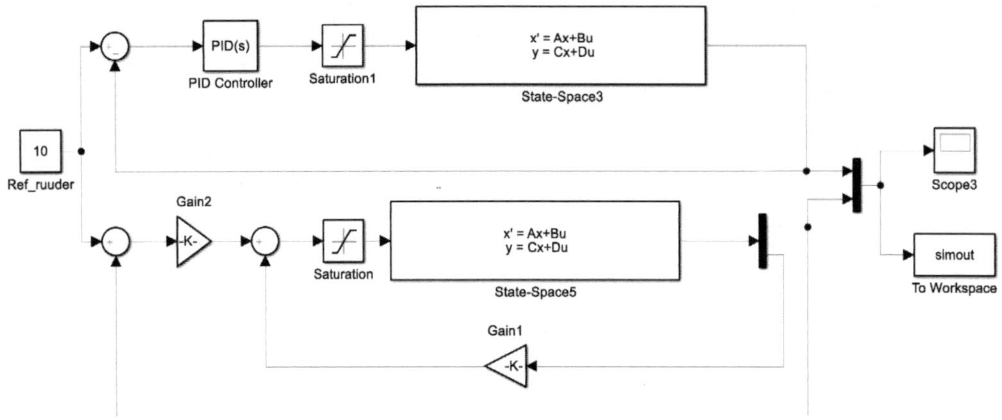

Figure 5. Simulink block diagram for LQR and PD based control design

4. EXPERIMENTAL SETUP

4.1 SINGLE BOARD PROGRAMMABLE CONTROLLER

The main electronics used in the setup is shown in Figure 6, the details description is given in [15, 16]. Single board programmable controller (Refer Figure 6 (a)) is the heart of the entire control and command system and generates computed control signals for control of the main propulsion motor as well as the stepper motor to operate the twin rudders. It also interfaces with the motion measurement and heading angle sensor unit i.e., the Motion Reference Unit (MRU). A 4-channel 24V relay module is the interface between the propulsion motor and the Electronic Speed Controller (ESC) and also for the steering stepper motor and data acquisition.

(a)

(b)

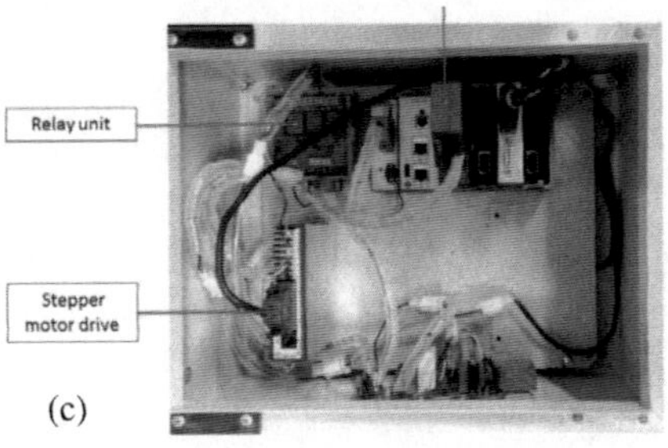

(c)

Figure 6. Embedded System: (a) Main controller (b) Digital I/O module,, (c) Suitable electronics

4.2 DATA ACQUISITION SYSTEM

The roll, pitch, heave and heading are measured electronically using a Motion Reference Unit (MRU). The MRU is placed at the center of gravity of loaded ship model. The MRU data is stored in a USB flash drive on board and at the same time data is transferred from the ship model to the base station through a wireless communication. The two-way data acquisition storage i.e. via USB and Wi-Fi gives data redundancy for safety. Secondly when data from multiple controlled platforms is required for group control, this is a minimum requirement.

4.3 PROPULSION SYSTEM

The main propulsion system consists of a 440 watt BLDC motor, a pair of transmission shafts, 1:1 counter-rotating gearbox and twin skewed five bladed propellers. The BLDC motor is interfaced with the main controller through an electronic motor drive. The drive gets the control signals from the main controller through a switching relay circuit. The RPM of the BLDC motor is controlled through the analog voltage generated by the controller. These analog voltage values can be changed using a virtual knob placed on a Graphical User Interface (GUI) window on the base station.

4.4 TWIN RUDDER STEERING SYSTEM

The steering system consists of a twin rudder installed with a stepper motor with the help of timing belt and pulleys. A stepper motor actuates the rudders with a holding torque capacity of 1.26N-m. The stepper motor is interfaced with the main controller through a stepper drive. The stepper motor operates in open loop as well as in closed loop. In open loop the rudder deflection commands are given by the user from the base station over wireless communication to carry out course correction in real time. In closed loop the system is set to follow the desired path where the steering control takes real time position data from the MRU and the control law generates the actuator signal to perform heading correction to keep the desired course. In the event of external disturbances only closed loop control will work.

4.5 COMMUNICATION SYSTEM

The on board system and the base station computer communicates over wireless link. This end-to-end communication uses client- server model, where the onboard system works as server, which share its resources to the base station, which acts as a client. This architecture provides the operating system independency and scalability of the whole setup. A wireless dual band router is used for the wireless communication. The router uses the latest high-speed wireless technology to bring lightning-fast Wi-Fi speeds of up to 433 Mbps on the 5 GHz frequency band and 300 Mbps on the 2.4 GHz frequency band.

5.5 POWER

All the components onboard are powered by three sets of lead acid batteries.

i. 12 Volt 26Ah batteries: A set of 4 batteries is connected in series to power the BLDC motor.
ii. 12 Volt 7Ah batteries: A set of 4 batteries is connected in series which powers the Stepper motor, Controller unit and MRU.
iii. 12 Volt 7Ah batteries: A single battery is used to power the Wi-Fi Router.

Figure 7. Model of Autonomous Ship in wave basin during maneuvering tests

The illustration photographs in Figure 7 amply demonstrate the versatility achieved in free model running.

6 CONCLUSIONS

This paper reports the development of a state-of-the-art full free self-propulsion maneuvering simulation system in a laboratory environment using robust, hi-fidelity wireless enabled Wi-Fi based communication protocol, enables IOT implementation and extendable to autonomous group control of surface ships.

9. REFERENCES

1. KRZYSZTOF WRÓBEL, JAKUB MONTEWKA, PENTTI KUJALA,Towards the assessment of potential impact of unmanned vessels on maritime transportation safety, *Reliability Engineering & System Safety*,Volume 165, 2017,Pages 155-169.

2. BURMEISTER, H.C.; PORATHE, T.; RØDSETH, Ø. J. (2013) Maritime Unmanned Navigation through Intelligence in Net- works: The MUNIN project, *12th International Conference on Computer and IT Applications in the Maritime Industries, COMPIT13, Cortona,* pp. 177-183

3. BRANDSÆTER, A.; TVETE, H. A. (2015), *Revolt; the unmanned, zero emission, short sea ship of the future*, Electric and Hybrid Marine World Expo, Amsterdam

4. ROLLS-ROYCE (2016), (Accessed 12/09/2016), *Remote and autonomous ships, the next steps*, http://www.rolls-royce.com

5. BURMEISTER, H.C.; PORATHE, T.; RØDSETH, Ø. J. (2013) *Maritime Unmanned Navigation through Intelligence in Net- works: The MUNIN project,* 12th International Conference on Computer and IT Applications in the Maritime Industries, COMPIT13, *Cortona,* pp. 177-183

6. NEGENBORN, R. (2014), *Inter-terminal transport on Maasvlakte 1 and 2 in 2030*, Delft University of Technology

7. BRANDS, M. (2015), *A Simio simulation model for the evaluation of Inter Terminal Transport systems at Maasvlakte 1 and 2 in 2030*, Delft University of Technology,

8. LODEWIJKS, G.; NEGENBORN, R.; ZHENG, H. (2015). *Predictive path following with arrival time awareness for water- borne AGVs*, Transportation Research Part C: Emerging Technologies

9. FOSSEN , T. I., Handbook of marine craft hydrodynamics and motion control (1994).

10. SKJETNE, R., SMOGELI, Ø. N., and FOSSEN, T. I. A nonlinear ship manoeuvering model:

Identification and adaptive control with experiments for a model ship, (2004). pp. 3-27

11. ÅSTRÖM, K. J., and KÄLLSTRÖM, C. G. (1976). Identification of ship steering dynamics. *Automatica*, ***12***(1), pp. 9-22..

12. QIN, L. Z. Y. Parametric identification of ships maneuvering motion based on Kalman filter algorithm, Mechatronics and Automatic Control Systems, Springer, 2014, 107-114.

13. PERERA, L. P. , OLIVEIRA P. and SOARES C. G. , System identification of nonlinear vessel steering, Journal of Offshore Mechanics and Arctic Engineering, 2015, **137**(3), 79-88.

14. NOMOTO, K. and TAGUCHI, K. H. S., On the Steering Quality of Ships, International Shipbuilding Progress, 1958, **4**(2), 354-370.

15. AWANISH C DUBEY, V. ANANTHA SUBRAMANIAN, V. JAGADEESH KUMAR Embedded system design for autonomous Unmanned Surface Vehicles. *4th International Conference in Ocean Engineering, 18-21 February,2018*

16. AWANISH C. DUBEY, NAGA VENKATA RAKESH, V. ANANTHA SUBRAMANIAN, JAGADEESH KUMAR V, . Wi-Fi Enabled Autonomous Ship Model Tests For Ship Motion Dynamics And Seakeeping Assessment. *IIRE Journal of Maritime Research and Development*, **1**(2) ISSN: 2456-7035

10. AUTHORS BIOGRAPHY

Brief biographies are required for all authors e.g.:

Awanish Chandra Dubey received his MTech degree from Devi Ahilya Vishwavidyalaya University in 2011. From 2011 to 2013, he was a research assistant in National Institute of Oceanography NIO Goa. He is currently a PhD scholar at The Department of Ocean Engineering, IIT madras. His research interest includes control of unmanned marine vehicles, underwater navigation and path planning. He is a co-inventor for the patent (EP2876520 A1), which is on "An altitude controlled profiler for use in coastal waters". He is a co-inventor for Wi-Fi based fleet control, system for application on board remotely operated surface vehicle.

V.Anantha Subramanian, Professor, has been with the Department of Ocean Engineering, IIT Madras for the last 35 years. His major area of specialization is naval architecture and Ocean Engineering. His interests are computational fluid dynamics applications related to ships and systems, computer aided ship design, practical ship hydrodynamics and testing and optimization related to ships and floating bodies. As a naval architect, he has been successful consultant to the shipbuilding and design industry catering to the requirements of clients both nationally as well as abroad.

P. Krishnankutty is a Professor in the Department of Ocean Engineering, Indian Institute of Technology (IIT) Madras. He works in the areas of ship motion and manoeuvring, powering and propulsion, and wave-structure interaction. He is a fellow of The Royal Institution of Naval Architects.

CHALLENGES OF MISSION PLANNING FOR COLLABORATIVE MARITIME AUTONOMY

S.R. Turnock, T.S. Hoang, S.J. Johnston, J. Downes University of Southampton, UK
J. Lam, and B. Pritchard, Thales Maritime Autonomy Centre, UK

SUMMARY

It is expected that a key part of a future, with ever increasing use of, maritime autonomous systems will require collaborative and integrated working. How approaches to such collaborative working can be developed and tested was part of the Integrated Mission Management System 2019 (IMMS2019) project. The immediate objective was to demonstrate the ability to use a single mission management system to control a fleet of heterogeneous autonomous platforms whereby their collaborative mission could be planned, verified and delivered. In order to challenge the various spatial-temporal-energetic-communication-environmental constraints when operating such a fleet, real-world demonstration trials were carried out in Plymouth sound deployed from the Thales Maritime Autonomy Centre at Turnchapel Wharf. The trials were focussed on ensuring that the control of the mission was secure and robust and that the need for human intervention was minimised. The trials were based on use of a 12m Rigid-Hulled Inflatable Boat (RHIB) that acted as the mobile command centre, a C-CAT3 Autonomous Surface Vessel (ASV), a fixed wing autonomous aircraft, and a pair of EcoSub Autonomous Underwater Vehicles (AUV). The successful completion of the trials provided essential information into how such collaborations can be best executed, alongside invaluable lessons as to how existing platforms need to enhance their interoperability and in particular, robustness of communications and mission plans.

NOMENCLATURE

AIS	Automatic Identification System
ASV	Autonomous Surface Vessel
AUV	Autonomous Underwater Vehicle
IMMS	Integrated Mission Management System
MAS	Maritime Autonomous System
NOC	National Oceanography Centre
RHIB	Rigid-Hulled Inflatable Boat
UAV	Unmanned Aerial Vehicle
UKHO	UK Hydrographic Office
UoS	University of Southampton
USV	Refer to ASV
UUV	Refer to AUV
UxV	Generic unspecified Unmanned Vehicle or Autonomous Vehicle i.e. Unmanned <<insert>> Vehicle

1. INTRODUCTION

The prospect of autonomous ships has received significant interest in recent years and, in particular with an appropriate focus on exactly how such systems can be safely operated in an assured regulatory framework. As with the on-going development of autonomous road vehicles, the degree of autonomy allowed to such a vehicle will require a progressive development process ensuring confidence in each successive phase. In this sense it is more appropriate to use the term Maritime Autonomous Systems (MAS) rather than ship as the public's perception of an autonomous ship as a crewless, independent vessel self-navigating (with little or no human intervention) from port-to-port, remains for a distant future. However, collective behaviours of heterogeneous systems will become ever more prevalent and it is important to understand and gather more detailed knowledge of these challenges of such interactions during actual maritime environments that inspired a joint project between Thales UK and University of Southampton (UoS) in early 2019.

The specific aim of the project was to investigate the challenges in the collaborative activity of maritime autonomous systems. This was achieved by conducting a series of field trials which examined how best to plan and execute missions at sea for three heterogeneous assets. The goal was to investigate how such a mission could be executed in such a way that the number of people involved can be significantly reduced. For the majority of MAS working today, it is still true that there is an inverted pyramid of effort required e.g. the number of people required is significantly larger than the number of MAS. This clearly highlights the difficulties of a crewless autonomous ship. If for example, it takes a team of four or more to operate a small displacement ASV from a science vessel, then scaling up to a collaborative environment of many MAS implies a significant cost both in terms of operational as well as human resources.

The later sections describe the implemented trials programme conducted in Plymouth Sound in October 2019 and the lessons learnt. Section 2 gives background to issues associated with MAS.

2. MARITIME AUTONOMOUS SYSTEMS

MAS, such as example assets of University of Southampton as shown in Fig. 1, offer the potential to reduce costs and remove humans from harm's way. The seas and oceans present formidable challenges to the adoption of autonomous systems and, in particular the collaborative co-ordination of activity of multiple platforms required to operate across significant spatial and temporal scales. Energy restrictions force choices between speed of transit and mission duration. Similarly ensuring the resilience of complex multi-component systems is vital to allow successful completion of each individual task which inevitably will require systems that can adapt to individual component failure, changing conditions or mission alterations.

Today's ocean exploration and underwater operations, in the majority of cases, requires the presence of a large dedicated vessel such as the Southampton-based RRS James Cook. Research cruises involve substantial crews of sailors, scientists and technicians. Although an increasing use is being made of AUVs such as the Autosub Long Range, it is still rare to attempt to coordinate the activities of multiples of such vehicles. Typically, these missions have a standard duration of ~8 to 24 hours, with the research ship needed to deploy, escort, monitor and recover these vehicles. The most detailed seabed survey to date using multiple assets was that by Thornton *et al* [1] on behalf of the Schmidt Institute in the summer of 2018 when 11.8 hectares of seafloor were surveyed [2]. The oil and gas industry uses AUV for surveys of underwater assets such as pipelines. Often the vessel 'follows' the track of the AUV (speed 1 to 2 m/s) on the surface to provide a navigational position reference and to allow some communication when faults are found.

(a) Maritime trials using SPOTTER aircraft

(b) L3-Harris ASV C-CAT3 platform

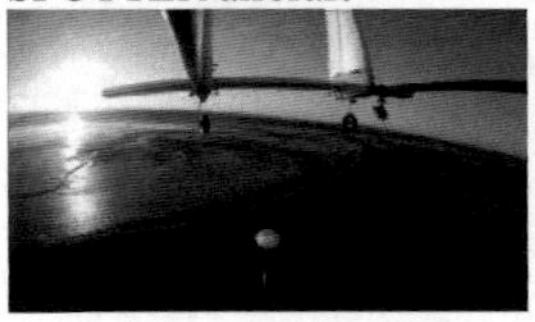

(c)Air drop testing of Eco-sub from SPOTTER aircraft at Thorney Island

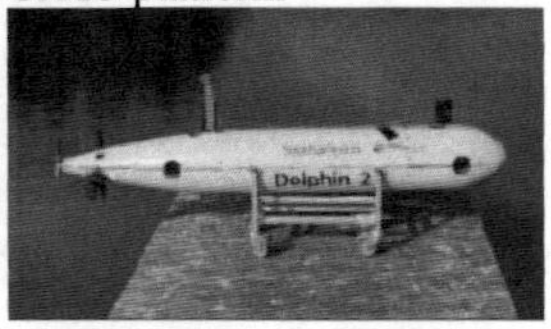

(d) Delphin 2 Autonomous Underwater Vehicle prepared for a lake trial

Figure 1 Four examples of low cost University of Southampton MAS assets

Long duration missions are possible with underwater glider AUVs (speed ~0.5 m/s) which have traversed the Pacific over several months and use a buoyancy engine to provide a saw tooth glide motion to sample the ocean environment. A next generation of glider [3] is under development in the Bridges H2020 project, which will go deep (~2000m +) and include a propeller so they can stay at depth and survey the seabed. However, there is limited ability to coordinate such activity due to communication limited to when these gliders surface and energy/costs limits with the use of satellite communications. The gliders use dead reckoning and a depth sensor to provide a measurement location. Sensors on gliders complement the vital international network of nearly 4000 Argo floats [4] which continually ascend/descend while they drift passively on the ocean currents.

The advances in MAS mirrors that across many domains such as driverless cars or drones. The development of robust and reliable platforms has proved challenging and organisations such as UK's National Oceanography Centre (NOC) have been working on its Autosub suite of vehicles for over 30 years, but still has only a limited number to deploy as each new vehicle requires a multi-million pound investment. Such valuable assets result in cautious approaches to AUV deployments and operations to minimise risk of losses [5].

Developments in ASVs, whose performance is less limited by stored energy, navigational and communication constraints have been more rapid, with for example Thales having taken delivery in 2018 from L3 Harris ASV of a 12m anti-mine system [6] capable of deploying aerial drones and underwater systems. PlanetOcean, in collaboration with NOC, L3 Harris ASV and UoS, funded by InnovateUK, have developed the Ecosub [7]. This smaller, 0.5 m, version of Autosub that can be air launched from the Spotter autonomous aircraft Fig. 1(a) or deployed from an ASV to provide a short duration survey.

The current state-of-the-art[8] either relies on a large inverted pyramid of effort to deploy a small number of autonomous platforms or as with the Argo floats and gliders allows only very limited control as to their activity or location. These limitations are driven by the nature of the marine environment. Achieving a reasonable speed requires a ship or submarine to be large which in turns requires substantial stored energy. Localisation of position underwater is limited by either the ability to discern terrain features or the existence of a nearby baseline acoustic reference.

There are two obvious routes that will significantly reduce the cost of exploring the ocean. First, is to provide a simple means of getting multiple MAS of differing design and domain (heterogeneous) to collaborate without needing a stovepipe of many individuals to operate each vehicle as was required in Unmanned Warrior 2016 [9]. Second, is to use the collaborative ability of many such MAS so that they can be projected across a wide geographical range, for example using a fast ASV or an autonomous aircraft. The collective assembly of individual MAS can then provide a method for localisation, can relay messages, interact with human supervisors and adapt their behaviour based on their local sensors. Scaling up from a small number of different MAS to one where there are many of varying type requires the use of algorithms that can provide assurance that the best choices are being made and that the team of human operators are not overwhelmed [10]. A key risk with autonomous systems is the trustworthiness of the decision-making and control mechanisms that substitute for or supporthuman control: to be trustworthy, systems need to remain safe while being resilient to unpredictable changes, mechanism failures and cybersecurity threats [11].

3. IMMS 2019 PROJECT

The Integrated Mission Management System 2019 (IMMS2019) project set out to create an integrated mission planning system for a heterogeneous multi domain fleet of MAS. The goals of this were to "establish a low-cost multi vehicle experimentation capability" through the creation of a joint team between Thales and the UoS and generating relevant research questions for future joint research.

The Integrated Mission Management System (IMMS) is an overarching architecture framework that allows the interoperability of diverse sets of autonomous systems operating in the sub-surface, surface, and aerial domains. At its heart is the IMMS software which manages the interfaces between a human operator(s) and the autonomous assets. The IMMS performs global planning and provides a mission definition across multiple platforms, while enabling an operator(s) to monitor the mission during its execution. This is commonly termed a Tier 1 level mission planner.

This system was designed to investigate improvements in the following performance metrics:

(i) **Scalability** – Reducing the people-to-asset ratio,
(ii) **Reusability** – Expanding asset functionality and ease of Plug and Play,
(ii) **Flexibility** – Speeding up and simplifying mission planning, re-planning and execution.

The focus of the project was a set of sea trials to test the IMMS with real assets in a representative environment, generating a comprehensive data set covering all aspects of the system. In addition, the act of planning and executing a trials event facilitated the identification of research questions in the field of integrated mission management of MAS.

For the purposes of this project, a use case scenario was proposed by Thales UK's Maritime Mission Systems business unit and further developed jointly within the project to test the development of an IMMS.

The scenario context used is to conduct a timely survey of the seabed in a defined area using available UxV assets. In addition to the base station or support ship, there is at least one available autonomous vessel for each of the aerial, surface and underwater domains. Fig. 2 illustrates the scenario in which the IMMS is based on the support ship. During the trials, the method of communicating with the AUVs was modified as shown in Fig. 3 where the AUV is connected directly to the IMMS when surfacing rather than using acoustic methods.

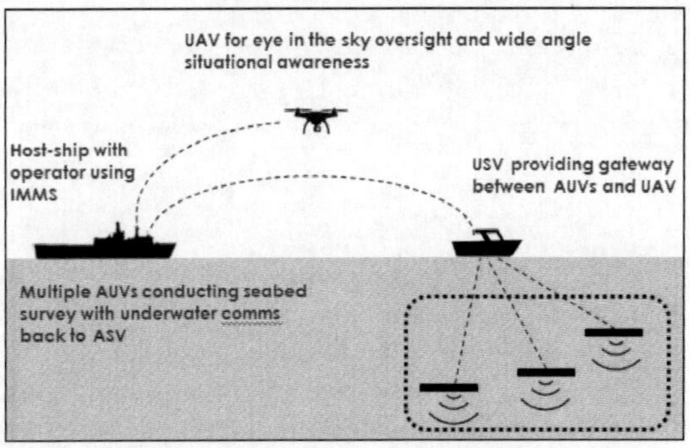

Figure 2 Original proposed use case scenario

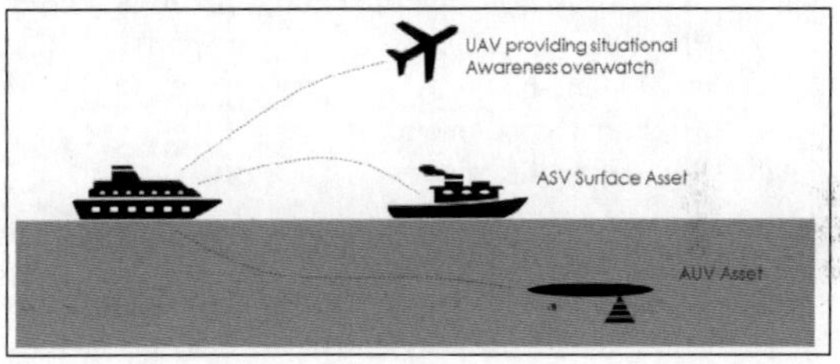

Figure 3 Use case as tested in sea trials

4. IMMS SYSTEM DESIGN

The IMMS system developed for this trial was predominately software focussed but also had key aspects of physical interfaces to connect to the autonomous vessels. The key functions of the trial for the IMMS 2019 system have been grouped as follows, to have:

- A **user interface** for creating/planning a mission, previewing a mission, live monitoring (situational awareness of live assets) and recording.
- **Path Planning** and **Optimisation** including **Validation and Verification** of Mission Plans
- **Conversion** of **"Approved" mission plans** to the individual trial assets. The overview of the process is shown in Fig. 4.

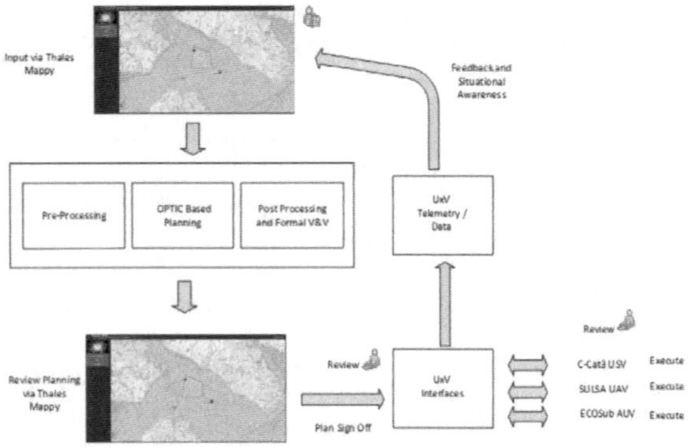

Figure 4 IMMS 2019 Conceptual Flow Diagram

The early design framework of developing the IMMS for heterogeneous autonomous assets is described in Dghaym *et al* [12]. The mission management activities required of the IMMS and the operator(s) involves:

(i) Planning of a mission after identifying the mission goals,
(ii) Mission execution and,
(iii) Reviewing the mission.

The need to have individually trustworthy autonomous vehicles working as separate entities is supplemented by the need to build a trustworthy management system so to ensure the trustworthiness of the overall system. Studies have shown that the cost of fixing errors during testing is ten times more than during the construction phase and can increase to more than 25 times post release [13], and many problems discovered in software systems are related to shortcomings in requirements elicitation and specification [14]. Formal modelling using *EventB* was applied to develop a requirements analysis framework for identification of anticipated range of operational environments for autonomous missions, including human operator interactions, together with precise specification of safety and security envelopes for enactment of autonomous missions. This framework was used to validate and verify the mission plan for all the assets for a given mission

5. TRIALS PREPARATION

Trials were conducted from Thales UK's maritime integration facility at Turnchapel Wharf, Plymouth. In terms of resources, the UoS staff provided an autonomous surface vessel, submarine and aerial vehicle and associated equipment for the trials. Thales provided the Turnchapel Wharf facilities, associated support vessel, and highly supportive staff. The joint team developed the IMMS.

5.1 Facilities

The facilities made available for the trials included enclosed hanger space (Fig. 5) and the RHIB 'Clyde of Dartmouth' as shown in Fig. 6, and access to the water via the Turnchapel Wharf slipway.

(a) Birds eye view (b) Hanger

Figure 5 Facility at Turnchapel Wharf, Plymouth

Figure 6 RHIB 'Clyde of Dartmouth' at Turnchapel Wharf

The hangar was used to store equipment and assets, and as a space to set up a workspace and safely recharge batteries. The RHIB provided the primary on water support and housed the developed IMMS system. Adaptations were necessary to enable the RHIB to fully support the trials, as the systems placed on board for the trials required significant electrical power. The RHIB was equipped with two 230Ah of auxiliary batteries which were able to charge from the engines providing via a 600W inverter to two standard power sockets.

5.2 Trials Area

A suitable trials area and UAV launch/landing site had been identified within the trials preparations previously on the other side of Plymouth Sound in Cawsand Bay (Fig. 7). This area provided sufficient water depth for the AUV to use, was free from regular marine traffic, shielded (to an extent) from adverse weather which typically prevailed from the south west, and contained an area considered suitable for aerial operations.

Aerial operations were conducted from the East Cornwall Scout Hooe Lake Campsite for which appropriate permissions for aerial operations were obtained. This is marked in Fig. 7 as 'SULSA Launch site' and shown in Fig. 8.

Figure 8 Aerial Platform Operating Site at Hooe Lake Campsite

5.3 MAS Trials Assets

Four assets were used during the Trials with overview specifications given in Tables 1 to 4.

Table 1 AUV Ecosub Robotics EcoSub µ5

Principal Particulars:	
Length (including antenna)	925mm
Diameter	111mm
Weight in Air	4Kg inc Batteries
Depth Rating	500m
Battery Technology	Manganese Alkaline
Speed	1m/s
Range	40Km
Endurance	Upto 12 hours
Communications (surface)	Iridium SBD, WiFi, GPS, Infrared and visible beacons
Communications (submerged)	Acoustic (optional)
Housekeeping	Internal temperature, pressure and humidity
CTD Sensor	NOC Micro-CT sensor (Conductivity and Temperature) (not commercially available)

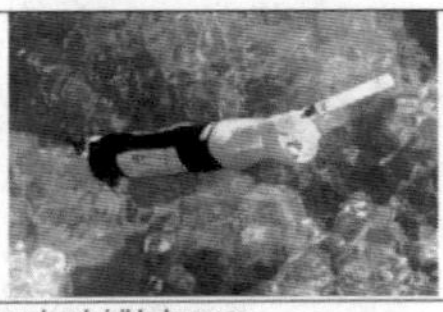

Table 2 ASV L3 Harris C-CAT3

Principal Particulars:	
Length	3.02m
Beam	1.55m
Height	1.4m
Draft	Min = 0.39m Max (Payload depending) ~ 0.7m
Displacement	Lightship 270Kg Full Load 350Kg
Primary Propulsion	2 x DC electric motors driving 3 bladed propellers
Operational Speed Range	Up to 8 knots
Payload Power	24V
Sensor Fit:	HD Fwd Facing Camera AIS Receiver IMS (not part of ASView sensor suite)
Payload Fit:	Independent Payload Computer for ROS interface

Table 3 UAV University of Southampton SULSA

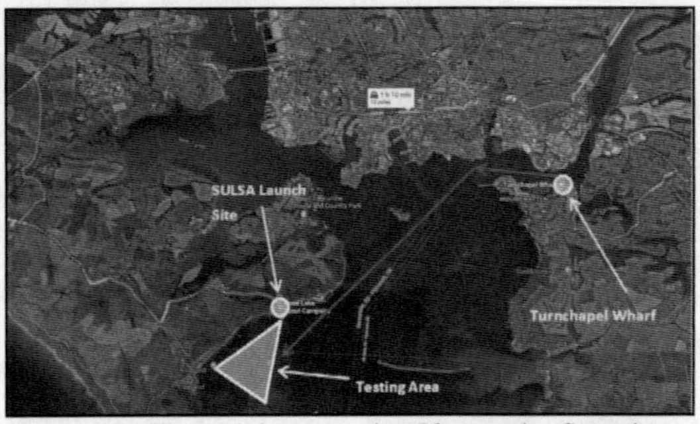

Figure 7 The trials area in Plymouth Sound was constrained within the coordinates [050.340, -004.181], [050.328, -004.185], [050.333, -004.197].

Principal Particulars:	
Wingspan	1.2m
Cruise Speed	60Knts
Max Take-off Weight	4Kg
Power Plant	1 brushless electric motor
Autopilot	ArduPilot
Payload	2 HD wide angle video cameras – 1 fwd facing, 1 facing to the left hand side. DTC Sol7 video transmission system
The World's first 3D printed aircraft	

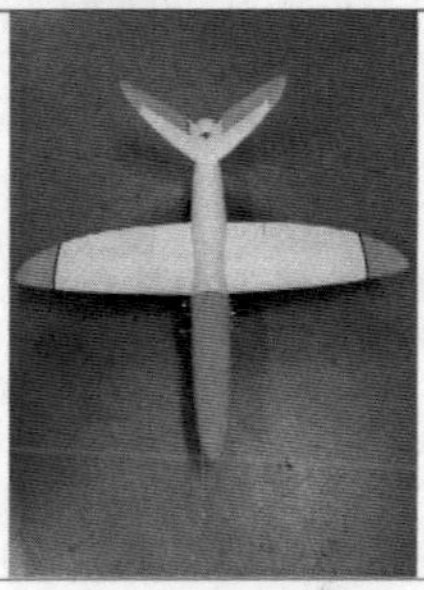

Table 4 UAV Avios Bushmule

Principal Particulars:	
Wingspan	1.5m
Length	1.095m
Cruise Speed	~30knts
Max Take-off Weight	2.25Kg
Power Plant	Twin electric 850KV brushless outrunner motors
Payload	2 HD wide angle video cameras – 1 fwd facing, 1 facing to the left hand side. DTC Sol7 video transmission system

5.4 Pre-trial approval process

To ensure that the trials were delivered in a safe manner that minimises risk to life and the assets, whilst enabling suitable research endeavour to be undertaken, there were considerable pre-trial planning and approval processes conducted leading up to the trials week. In compliance with the Queens Harbour Masters requirements for Plymouth and the Maritime UK Industry Code of Practice, the Turnchapel Wharf Trials Manager distributed a notice to mariners warning of the trials planned for this location. The area selected for trials (Cawsand Bay) is a designated area for conducting trials in Plymouth sounds.

The UAV launch/landing location was identified on Google Maps satellite view, surveyed by the Turnchapel team and was confirmed to meet the requirements specified by the UAV operations team of 50m of flat grass and accessibility for the mission control van. A CAA Standard Permission to Operate Unmanned

Aircraft Systems in UK Airspace was applied for and successfully obtained by University of Southampton SOTON UAV team for the trials.

A combined safety and risk assessment process was developed by the project team that satisfies the requirements of both Thales and the UoS. The overall process is illustrated in Fig. 9. As can be understood this required a significant investment of time and is worth bearing in mind as the prospect of such combined operations becomes more prevalent.

A detailed trials plan was developed. This plan covered the aims and objectives of the trials; any pre-requisites; details on the location, equipment and resources; relevant regulations; limits of operating conditions; detailed daily schedule; Personal Protective Equipment (PPE); and the safety & risk assessment descriptions.

The trials plan was supplemented by a detailed trials playbook, produced prior to the trials. The described activities consisted of basic safety tests for the assets, tests to evaluate various functionalities from the original IMMS requirements, and demonstrations of the IMMS in realistic use case scenarios together with the assets. These activities were divided across the trials week which are ordered with increasing complexity throughout the week. It was intended that it would therefore be known which tasks the IMMS and assets would be capable of attempting, and the day's activities could be defined accordingly in a briefing session each morning. The playbook described more activity items than were feasible in a single day's trials. The purpose of this was to allow the playbook to be adaptable to suit changing environmental conditions and unknown levels of progress across the week.

A window of time was identified with predicted weather forecast conditions used to determine the final choice. The final trials took place in early October and experienced a variety of weather conditions which did impact on the time available each day for testing. The decision to go ahead was based on a Met Office long range weather forecast:

Thursday 3rd Oct 2019 - Saturday 12th Oct 2019
"It will be a chilly start on Thursday, with a rural air frost. Outbreaks of rain and strong winds will spread into western areas by the afternoon though, with gales possible. This unsettled weather will then move eastwards later in the day. By the weekend it is likely that unsettled, wet and possibly very windy weather will return, bringing with it a risk of gales, especially in the north and west. During the following week, the unsettled weather may begin to become more generally confined to the north and west, with occasional spells of drier and brighter weather, particularly in the south and east. Temperatures are likely to vary around average, although some cold nights are possible under any clear skies."

The UK Hydrographic Office (UKHO) made available detailed Hydrographic data covering the test area in Cawsand bay as well as historic Automatic Identification System (AIS) data for the Plymouth region. These were displayed in the mission interface. This proved useful as a sanity check during one mission, as it was realised the end point of the C-Cat mission was outside the bathymetric data, and the sea bed was too shallow to safely run trials in that area, despite the conventional charts and maps showing it as sea.

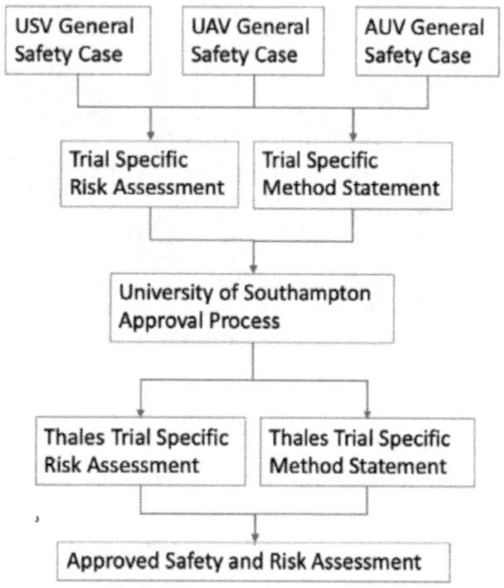

Figure 9 Safety and Risk Assessment Process

6. TRIALS

Over the five days of the trials (Table 5) a progressive series of trials developed confidence that all the necessary components of the system could function and could be deployed. There were a number of issues with individual assets and in particular communication problems were identified that required reference back to the supplier on a number of times. Key achievements were the ability for all assets and base stations including not just the RHIB but also the crew at Turnchapel and at the UAV control van to sharing the same mission information using a common system.

Table 5 Planned Trials schedule.

DAY 1 *Monday October 7th*	AM PM	Travel from Southampton/Reading Initial Platform Setup and Briefing
DAY 2 *Tuesday October 8th*		Single Platform IMMS Trials
DAY 3 *Wednesday October 9th*		Mixed Platform IMMS Trials
DAY 4 *Thursday October 10th*		Integrated Platforms Demonstration
DAY 5 *Friday October 11th*	AM PM	Trials Review Meeting Travel back to Southampton/Reading

The following provided edited highlights from the test logs during the trials week which capture many of the challenges that are present in real world testing of MAS.

Day 1

"The objective of activity for Monday according to the trials plan was dedicated entirely to system setup, covering the communications network on the Clyde and checking all UxVs to be operating properly. Time was also put aside to brief the full team on plans and safety aspects for the week as well as give team introductions and the Turnchapel site. The plans for Monday afternoon were not significantly altered. All communications equipment was mounted to Clyde and shown to be in working order. The UAV SULSA was able to interpret an example mission generated using the IMMS. No attempt was made to fly as the UAV team undertook a reconnaissance of the UAV airfield site having not previously flown from that site."

Day 2

"The goals on day 2 were to see the C-Cat 3 fully operational at Turnchapel and to perform initial test flights with the UAV SULSA. The weather was expected to be particularly poor on Tuesday morning making a trip to the test area less desirable. The seagoing vehicles did not leave the Turnchapel Wharf area. The EcoSubs were not able to be ballasted to reliably sit at the required 70°±5° to fully raise an antenna above the water. Therefore EcoSubs were not launched due to risk of loss. C-Cat 3 was successfully launched and brought alongside Clyde. It was driven remotely using the ASView-Helm manual controls and operated reliably within ~20m of Clyde. Beyond this range, communications issues were experienced..

In a late afternoon weather window, UAV SULSA flew from the Hooe Lake campsite under manual control and recorded approximately 15 minutes of HD video (Fig. 10). The aircraft performed well at all points until landing. The heavy landing was attributed to the required upgrades to the aircraft increasing its mass and requiring greater airspeeds. The wind direction also forced the landing to take place downhill which worsened the situation A lighter replacement aircraft capable of running the same operating system was arranged for the rest of the week."

Figure 10 Still image from UAV SULSA

Day 3

"The C-Cat 3 was launched successfully and transited via RHIB Clyde of Dartmouth to the Cawsand Bay trials area and further single asset testing was undertaken. It was driven remotely using the ASView-Helm manual controls and operated reliably within ~200m of Clyde. Beyond this range, communications issues were experienced. Due to the issues with communications on the C-Cat 3 it was decided to halt the trials in Cawsand bay earlier than planned and return to Turnchapel Wharf for further investigation.

Via discussion with the support team at L3 Harris the underlying issues with communications were identified and rectified.. It was a frustrating day with the assets due to the ongoing communications issues on the C-Cat 3 but with good success solving these late in the afternoon. The replacement airframe flew successfully and proved to be adept at handling the challenges of the Hooe Lake campsite."

Day 4

"Day 4's objective was to undertake integrated mixed platform IMMS based trials. This was intended to create and validate missions within the IMMS system and use an integrated mix of differing assets to achieve the specified mission. Based on the condensed playbook developed on Day 3, five specific missions were defined for evaluation on Day 4 of the trials. It was decided by the team that that the most important outcome was to demonstrate the IMMS operating with multiple assets in one mission as this is where the core novelty lies within this project. These activities were further planned to include every element available to a mission in as few missions as possible.

It was successfully shownthat an integrated mission for multiple assets could be developed, planned, and distributed using the established IMMS2019 system. The UAV successfully executed that mission (Fig. 11). Ongoing hardware issues prevented the ASV C-Cat3 and AUV Ecosubs from executing but the missions were delivered to the assets."

7. CONCLUSION

A low cost multi vehicle experimentation capability was successfully developed and executed within a four-month period using a joint team of Thales and UoS engineers. The completion of a set of trials of increasing complexity allowed further exploration of the actual challenges when moving between a planned mission, via a virtual simulation to actual delivery.

Figure 11 UAV 1st IMMS 2019 uploaded mission flight plan in grey and track of autonomous flight in green.

The IMMS2019 System was successfully shown to be able to plan and execute multiple domain and multiple asset missions. The common network approach adopted enabled multiple sites and locations to participate in the trials. A joint team of 16 people in Turnchapel Wharf successfully delivered the 2019 trials

Significant challenges were experienced with the assets working in the field. These were primarily related to the use of proprietary systems, which stressed the importance of understanding interfaces and connectivity of closed systems vs. open systems. Real world communications provided significant challenges with inconsistencies during system set up and were challenging to detect and fault find. There were problems with the selected site for UAV launch and recovery, dependent on weather conditions as well as issues of energy reserves to ensure suitable range could be achieved. This was partly due to the topography of Plymouth Sound with regard to access to suitable landing strips but also reflects the energy resource issue of low cost assets.

While it was true that many of these issues were mainly associated with the need to mature the assets prior to, rather than during, the trial their operations and likewise the software system development required the combined operation of many individual systems in advance of large-scale future integrated mission trials.

The IMMS2019 system was immature at this stage and needs further stress testing in advance of the field trials. Issues were found in the aspects of the code which had earlier lab trials of the complete system been undertaken, may have been identified. Ultimately, the IMMS demonstrates the importance of research, engineering design, testing of autonomous vehicles, albeit the assets are simple, to determine the real behaviours and embedded costs associated with conducting more complex MAS missions going into the future.

8. ACKNOWLEDGEMENTS

IMMS2019 team led by Dr Jon Downes and Ben Pritchard are grateful to the many others both at Turnchapel Marine Autonomy Centre and in Southampton who helped ensure the success of the IMMS trials. In particular, Dr D Dghaym, Prof M. Butler, Prof E. Rogers and Prof J. Scanlan. The assistance of the UK Hydrography office in providing access to relevant environmental data proved invaluable. Likewise the EPSRC capital funding that supported the assets deployed in the trial.

9. REFERENCES

[1] https://schmidtocean.org/cruise/adaptive-robotics-at-barkley-canyon-and-hydrate-ridge/

[2] https://www.southampton.ac.uk/smmi/news/2018/09/13-rapid-data-analysis-at-sea.page

[3] Lidtke, A., Turnock, S. and Downes, J. (2018) Hydrodynamic Design of Underwater Gliders Using k-k(L)-omega Reynolds Averaged Navier-Stokes Transition Model. IEEE Journal of Oceanic Engineering, 43 (2), 356-368. (doi:10.1109/JOE.2017.2733778).

[4] http://www.argo.ucsd.edu/

[5] https://www.bbc.co.uk/news/science-environment-43378290,

[6] https://www.asvglobal.com/brest-first-images-of-the-thales-mine-warfare-usv/

[7] https://www.ecosub.uk/news

[8] Thompson, F & Guihen, D, Review of mission planning for autonomous marine vehicles fleets, J. of Field Robotics, 2018, DOI:10.1002/rob.21819

[9] https://www.royalnavy.mod.uk/unmannedwarrior

[10] Delle Fave, F. M., Rogers, A., Xu, Z., Sukkarieh, S. and Jennings, N. R. (2012) Deploying the Max-Sum Algorithm for Coordination and Task Allocation of Unmanned Aerial Vehicles for Live Aerial Imagery Collection. In, Proceedings of ICRA, 14 - 18 May 2012, 469-476.

[11] W. Schwarting, J. Alonso-Mora, D. Rus. Planning and Decision-Making for Autonomous Vehicles, Annual Review Control, Robotics, and Autonomous Systems,1:1 2018.1.

[12] Dghaym, D., Turnock, S., Butler, M., Downes, J., Hoang, T.S., Pritchard, B., 2019. Developing a framework for trustworthy autonomous maritime systems, in: International Seminar on Safety and Security of Autonomous Vessels (ISSAV): ISSAV 2019. URL: https://eprints.soton.ac.uk/434545/.

[13] Leffingwell, D., 1997. Calculating the Return Investment from more Effective Requirements Management. American Programmer, 10,13–16.

[14] MacDonell, S.G., Min, K., Connor, A.M., 2014. Autonomous requirements specification processing using natural language processing. CoRR abs/1407.6. URL: http://arxiv.org/abs/1407.6099, arXiv:1407.6099.

9. AUTHOR'S BIOGRAPHY

Professor Stephen Turnock, FRINA, is Head of Civil, Maritime and Environmental Engineering at the University of Southampton. His interests are in decarbonisation, maritime autonomy and performance sport.

Dr Thai Son Hoang is a Lecturer in Cyber-Physical Systems at the University of Southampton. His research interests include formal methods and tools for developing safe and secure systems.

Ben Pritchard, is the Research Group Leader for Autonomous Systems at Thales UK. He led the work on the IMMS for Thales and continues to work to mature these concepts for maritime customers and in other Thales market sectors.

Joseph Lam, is a Principal Solutions Architect at Thales UK and was part of the IMMS team. He has a number of years of systems engineering experience across the Aviation, Maritime, and Ground Transportation domains. His research interests are in mission systems in autonomy, enterprise and systems architecture.

Dr Jon Downes directs the Maritime Robotics Laboratory at UoS. An academic with experience in structures, remote inspection and maritime autonomy he has led the development of ASV capability at UoS in particular the EPSEC funded CCAT3.

CONTROL SYSTEM DESIGN FOR AUTONOMOUS UNDERWATER TRANSPORTATION USING FOUR SEAPERCH HAUVS

F. U. Rehman, G. Thomas and **E. Anderlini,** University College London, UK
S. Khalid and **Z. Riaz,** Pakistan Navy Engineering College, Pakistan

SUMMARY

In this paper, four Seaperch HAUVs were considered rigidly connected to a payload via solid plate links. A nonlinear coupled dynamic model was developed using the Newton-Euler formulation including the inertia, Coriolis, hydrodynamic, hydrostatic and thrust terms. The hydrodynamic parameters were calculated using the semiempirical approach. The vehicles and the solid links have their buoyancy greater than weight, however, the payload was selected such that the whole system is neutrally buoyant. Time-domain motion simulation was initially carried out without implementing the control system to verify the motion response. A control system was then designed using the PID controllers. Low PID gains were analysed to be enough to reach the desired location without following a trajectory. A minimum snap trajectory was then generated, and the control system was applied to follow it. At the lower PID gains, the trajectory following had a high time lag and also the thrust forces which were applied to follow the trajectory had slackness. This was not acceptable as this resulted in the failure of the transportation mission. Therefore, the PID gains were increased to a reasonable range and the acceptable trajectory tracking was achieved.

NOMENCLATURE

O	Centre of body
BFF	Body-fixed frame
EFF	Earth-fixed frame
COG	Centre of gravity
COB	Centre of buoyancy
$\boldsymbol{R}$	Translational transformation matrix
$\boldsymbol{T}$	Rotational transformation matrix
$\boldsymbol{J}$	Complete transformation matrix
m	Mass of system
(Ix, Iy, Iy)	Moment of inertia about x, y and z-axes
(X, Y, Z)	Surge, sway and heave force
(K, M, N)	Roll, pitch and yaw moments
(x, y, z)	Surge, sway and heave displacements
(ϕ, θ, ψ)	Roll, pitch and yaw angles
(u, v, w)	Surge, sway and heave velocities
(p, q, r)	Roll, pitch and yaw velocities
$\boldsymbol{M_{RB}}$	Rigid body mass matrix
$\boldsymbol{M_A}$	Added mass matrix
$\boldsymbol{M}$	Mass matrix of the system
$\boldsymbol{C_{RB}}$	Rigid body Coriolis matrix
$\boldsymbol{C_A(v)}$	Added mass Coriolis matrix
$\boldsymbol{C(v)}$	Combined Coriolis matrix
$\boldsymbol{D(v)}$	Damping matrix
$\boldsymbol{g(\eta)}$	Vector of hydrostatic parameters
$\boldsymbol{\tau}$	Thrust vector
$\boldsymbol{v}$	velocity vector
W	weight
B	Buoyancy
$(X_{\dot{u}}, Y_{\dot{v}}, Z_{\dot{w}})$	Translational added mass terms
$(K_{\dot{p}}, M_{\dot{q}}, N_{\dot{r}})$	Rotational added mass terms
$(X_{u\|u\|}, Y_{v\|v\|}, Z_{w\|w\|})$	Translational damping terms
$(K_{p\|p\|}, M_{q\|q\|}, N_{r\|r\|})$	Rotational damping terms
(τ_X, τ_Y, τ_Z)	Translational Actuator terms
(τ_K, τ_M, τ_N)	Rotational Actuator terms
$\boldsymbol{T_a}$	Combined thrust allocation matrix
$\boldsymbol{T}_i \; i = 1,..4$	Thrust allocation matrix of HAUV
K_P	Proportional gain
K_i	Integral gain
K_d	Differential gain
$\boldsymbol{\eta_d}$	Vector of desired state of the system
$\boldsymbol{\eta}$	Vector of actual state of the system
$\boldsymbol{e}$	Error between $\boldsymbol{\eta_d}$ and $\boldsymbol{\eta}$
$\boldsymbol{\tau_e}$	Thrust vector in EFF
$\boldsymbol{\tau_d}$	Thrust vector in BFF
$\boldsymbol{f_d}$	Desired thrust force vector in BFF
$\boldsymbol{f}$	Actual thrust force vector in BFF
$\boldsymbol{f_f}$	Vector of thrust force function
RPMs	Revolutions per minute

1. INTRODUCTION

Unmanned Underwater Vehicles (UUV) are increasingly used in the underwater operations due to its performance in circumstances which are impossible or difficult for humans [1]. It also avoids risking human lives [2]. Using multiple UUVs is advantageous as it enables lifting heavier loads, allows flexibility in deciding the number of vehicles and provides fault tolerance [3].

The two main types of UUV are Remotely Operated Vehicles (ROV) and Autonomous Underwater Vehicles (AUV). ROV has the advantage of slow precise operations. On the other hand, AUV is pre-programmed to be operated independently [4]. To get the advantage of both the ROV and AUV, an ROV can be modified to be used autonomously and is called the Hovering Autonomous Underwater Vehicle (HAUV) [5].

Though underwater transportation is not an efficient way of transportation due to the increased density and viscosity of water, however, it could be advantageous in the military field; by avoiding detection of military payload [4] and in the commercial field; by transporting and installing the oil and gas extraction rig [6].

There are two ways of transportation with multiple vehicles which are mostly used for the land and aerial vehicles. In the first method, the vehicles are considered rigidly connected to the payload via solid links, whereas, the second method treats the vehicles as they are connected flexibly to the payload through cables. For developing the simulation model of the rigid connection method, the whole system of vehicles connected to a payload is considered a single rigid body [7]. There is no margin for the motion of the individual vehicle in the system. On the other hand, the flexible connection method provides independency to the vehicles to maintain their positions w.r.t the payload [8].

Implementing the above methods of multi-vehicular transportation in the underwater environment is challenging. The significant terms in analysing the motion response of land and aerial vehicles are friction and gravity respectively [8]. The aerodynamic parameters are ignored due to their less significant values. On the other hand, underwater vehicles experience substantial hydrodynamic and hydrostatic parameters. Moreover, underwater dynamics is highly nonlinear and coupled.

In this paper, four Seaperch ROVs were modified to be used autonomously, therefore, are called the Seaperch HAUVs. The rigid connection method was used such that the HAUVs are rigidly connected to a payload made of a square plate via solid plate links. This system was selected for the simulations as the same is being fabricated in the lab for the validation of results in future.

2. DEVELOPMENT OF DYNAMIC MODEL

To analysis the motion response of the system of four Seaperch HAUVs, the dynamic model was developed using the approach derived by Thor I. Fossen [9].

Two reference frames were defined i.e. the earth-fixed and body-fixed frames, as shown in Figure 1. The position and orientation of the system are expressed in the earth-fixed frame, whereas, the velocity and force terms in the body-fixed frame.

The equations of motion were derived in the body-fixed frame by taking the effect of all the parameters about a combined centre of the body (O).

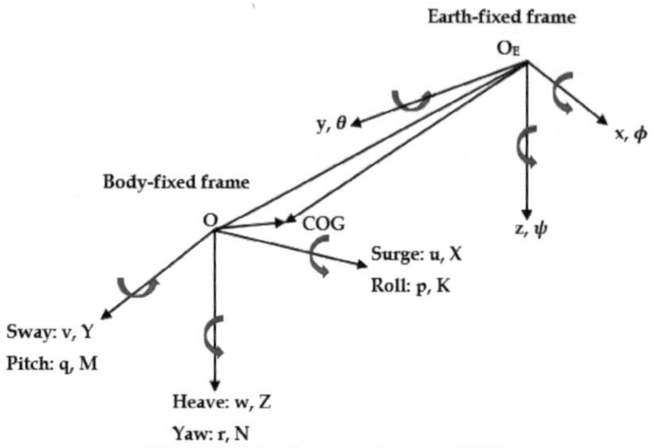

Figure 1: Reference frames [10]

2.1. Assumptions

Following assumptions were taken in the analysis

1. The system is a rigid body i.e. there is no relative motion between the mass particles when the force is applied [11].
2. The mass and its distribution do not change during the motion of the vehicle. This ensures that the weight, buoyancy, COG and COB do not vary during transportation.
3. The vehicle is deeply submerged; therefore, the wave effects are ignored.
4. The interaction effect is ignored.
5. Sea current effects are ignored.
6. The centre of the combined body is at the centre of the payload (O).
7. The combined system is neutrally buoyant.

2.2. Equations of Motion

The combined system has three planes of symmetry as the same Seaperch HAUVs are attached through the same solid links to a square plate (selected payload) at the positions which have the same distance from the centre of the payload, as shown in Figure 2. This helps in assuming that the COG and O are at the same position, and COB right above the COG in the vertical direction to ensure static stability of the system.

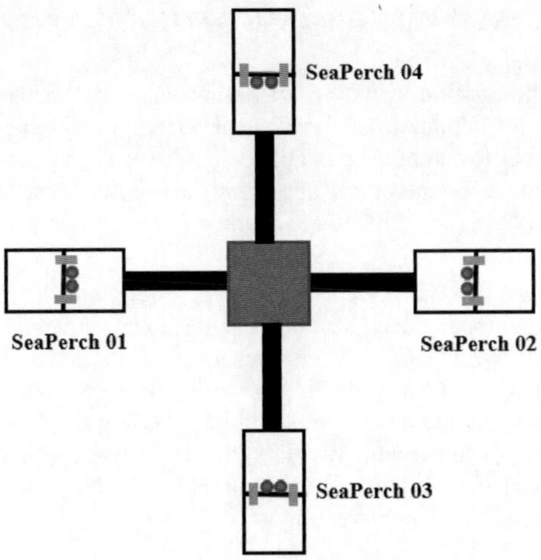

Figure 2: System setup

Including the hydrodynamic, hydrostatic, inertia and actuator terms of all the vehicles, links and payload, the combined equations of motion in 6 DOF are given as [10]

$$(m + X_{\dot{u}})\dot{u} + m(wq - vr) + Z_{\dot{w}}wq - Y_{\dot{v}}vr + X_{u|u|}u|u| + (W - B)sin\theta = \tau_X,$$

$$(m + Y_{\dot{v}})\dot{v} + m(ur - wp) - Z_{\dot{w}}wp + X_{\dot{u}}ur + Y_{v|v|}v|v| - (W - B)cos\theta sin\phi = \tau_Y,$$

$$(m + Z_{\dot{w}})\dot{w} + m(vp - uq) + Y_{\dot{v}}v - X_{\dot{u}}uq + Z_{w|w|}w|w| - (W - B)cos\theta cos\phi = \tau_Z,$$

$$(I_x + K_{\dot{p}})\dot{p} + (I_z - I_y)qr - (Y_{\dot{v}} - Z_{\dot{w}})vw - (M_{\dot{q}} - N_{\dot{r}})qr + K_{p|p|}p|p| - z_bBcos\theta sin\phi = \tau_K, \quad (1.\ 1)$$

$$(I_y + M_{\dot{q}})\dot{q} + (I_x - I_z)rp - (Z_{\dot{w}} - X_{\dot{u}})uw - (N_{\dot{r}} - K_{\dot{p}})pr + M_{q|q|}q|q| - z_bBsin\theta = \tau_M,$$

$$(I_z + N_{\dot{r}})\dot{r} + (I_y - I_x)pq - (X_{\dot{u}} - Y_{\dot{v}})uv - (K_{\dot{p}} - M_{\dot{q}})pq + N_{r|r|}r|r| = \tau_N.$$

Morison's equation was used to work out the hydrodynamic parameters which consist of the added mass and drag terms. The added mass part was derived based on the potential flow theory, whereas, the drag part was achieved from experiments [12]. The hydrodynamic coefficients of the vehicle, solid link and payload were calculated using the empirical data from the DNV standards [13], as shown in Table 1. The moment terms were taken about the centre of the combined body (O).

Seaperch ROV was made of pipes of different sizes, elbows and T joints. Moreover, the axial and vertical thrusters were installed on the centre pipe at an equal distance from O respectively, as shown in Figure 2. In order to ensure neutral buoyancy, buoyancy patches were used. The hydrodynamic parameter of each of the seaperch HAUV was calculated separately and summed together to get the total effect of each vehicle about O.

The hydrodynamic parameters of the HAUV, solid link and payload are mentioned in Table 1.

Table 1: Hydrodynamic parameters

		Added mass terms	**Damping terms**
Seaperch HAUV	Surge	0.819 Kg	13.12 Kg/m
	Sway	2.35 Kg	21.2 Kg/m
	Heave	2.56 Kg	21.5 Kg/m
	Roll	0.26 Kgm	0.0038 Kgm^2
	Pitch	4.88 Kgm	0.0082 Kgm^2
	Yaw	11.7 Kgm	0.0087 Kgm^2
Link	Surge	0.42 Kg	0.054 Kg/m
	Sway	0.2 Kg	1.17 Kg/m
	Heave	0.04 Kg	2.9 Kg/m
	Roll	1.24e-6 Kgm	1.6e-7 Kgm^2
	Pitch	0.066 Kgm	5e-4 Kgm^2
	Yaw	4.8 Kgm	2.9e-4 Kgm^2
Payload	Surge	0.023 Kg	0.16 Kg/m
	Sway	0.064 Kg	0.58 Kg/m
	Heave	0.1 Kg	2.41 Kg/m
	Roll	6.3e-2 Kgm	4.1e-6 Kgm^2
	Pitch	2.4e-3 Kgm	7.4e-5 Kgm^2
	Yaw	5.9e-3 Kgm	1.35e-5 Kgm^2

The hydrostatic parameters were included for all the bodies. The Seaperch HAUVs and the solid links have their buoyancy greater than weight in order to bring them to the surface in case of an emergency. In an emergency, the vehicles will drop the payload and come to the sea surface along with the solid links. On the other hand, the weight of the payload is greater than its buoyancy. However, the combined system is neutrally buoyant.

Equation (1. 1) can be expressed in the matrices form as [9]

$$\boldsymbol{M\dot{v}} + \boldsymbol{C(v)v} + \boldsymbol{D(v)v} + \boldsymbol{g(\eta)} = \boldsymbol{\tau}. \quad (1.\ 2)$$

Where

$\boldsymbol{M} = \boldsymbol{M_{RB}} + \boldsymbol{M_A}.$
$\boldsymbol{C(v)} = \boldsymbol{C_{RB}(v)} + \boldsymbol{C_A(v)}.$

3. PROPULSION MODEL

To develop the propulsion model for the combined system of four Seaperch HAUVs, the thrust vector of each thruster was worked out using the following standard [14]

$$\tau_{t_i} = \begin{bmatrix} \boldsymbol{f} \\ \boldsymbol{l_{t_i}} \times \boldsymbol{f} \end{bmatrix} = \begin{bmatrix} f_x \\ f_y \\ f_z \\ l_yf_z - l_zf_y \\ l_zf_x - l_xf_z \\ l_xf_y - l_yf_x \end{bmatrix}. \quad (1.\ 3)$$

$\boldsymbol{f}$ is the thrust force vector of the thrusters and $\boldsymbol{l_{t_i}}$ is the distance of that thruster from O.

The system under study consists of eight axial and eight vertical thrusters. The combined thrust vector is written as a product of thrust allocation matrix ($\boldsymbol{T_a}$) and thrust force vector ($\boldsymbol{f}$)

$$\boldsymbol{\tau} = \boldsymbol{T_a f}. \tag{1. 4}$$

The thrust force vector of the combined system is written as

$$\boldsymbol{f} = [f_1 \ f_2 \ f_3 \dots . f_{16}]^T. \tag{1. 5}$$

The thrust allocation matrices of the individual vehicles in the combined system are given as

$$\boldsymbol{T_1} = \begin{bmatrix} 1 & 1 & 0 & 0 \\ 0 & 0 & 0 & 0 \\ 0 & 0 & 1 & 1 \\ 0 & 0 & 0.025 & -0.025 \\ 0 & 0 & 0.42 & 0.42 \\ -0.055 & 0.055 & 0 & 0 \end{bmatrix}. \tag{1. 6}$$

$$\boldsymbol{T_2} = \begin{bmatrix} 1 & 1 & 0 & 0 \\ 0 & 0 & 0 & 0 \\ 0 & 0 & 1 & 1 \\ 0 & 0 & 0.025 & -0.025 \\ 0 & 0 & -0.42 & -0.42 \\ -0.055 & 0.055 & 0 & 0 \end{bmatrix}. \tag{1. 7}$$

$$\boldsymbol{T_3} = \begin{bmatrix} 0 & 0 & 0 & 0 \\ 1 & 1 & 0 & 0 \\ 0 & 0 & 1 & 1 \\ 0 & 0 & 0.42 & 0.42 \\ 0 & 0 & 0.025 & -0.025 \\ -0.055 & 0.055 & 0 & 0 \end{bmatrix}. \tag{1. 8}$$

$$\boldsymbol{T_4} = \begin{bmatrix} 0 & 0 & 0 & 0 \\ 1 & 1 & 0 & 0 \\ 0 & 0 & 1 & 1 \\ 0 & 0 & -0.42 & -0.42 \\ 0 & 0 & 0.025 & -0.025 \\ -0.055 & 0.055 & 0 & 0 \end{bmatrix}. \tag{1. 9}$$

$\boldsymbol{T_a}$ is given as

$$\boldsymbol{T_a} = [\boldsymbol{T_1} \, \boldsymbol{T_2} \, \boldsymbol{T_3} \, \boldsymbol{T_4}]. \tag{1. 10}$$

Each column in the thrust allocation matrix of a Seaperch HAUV defines the effect of a particular thruster w.r.t O.

Though each vehicle is underactuated as no thruster is applied in the transverse direction, however, the combined system is fully actuated as the axial thrusters of the transversely placed vehicles act in the transverse direction.

The thrust force produced by the thrusters at different RPMs was worked out using the load cell in the towing tank of University College London (UCL).

4. SYSTEM RESPONSE WITHOUT CONTROL

The motion response was first analysed without the implementation of the control system by giving thrust input to the thrusters. This helps to verify the simulation model. Each vehicle applies a maximum thrust force of 1.6N in each direction as worked out experimentally.

Test 01: The Seaperch HAUVs were attached in the combined system such that the axial thrusters apply positively inward. Therefore, when all the axial thrusters on the four Seaperch HAUVs apply their positive thrust at an equal amount, there should not be any motion. In this test, the maximum thrust force of 0.8N was applied by the axial thrusters on all the four HAUVs. No motion was observed in any direction.

Test 02: In this test, the maximum thrust force was applied by all the vertical thrusters in the combined system. The system moves a distance of 7.4m in the vertical in 30secs with zero motion in other directions, as shown in Figure 3.

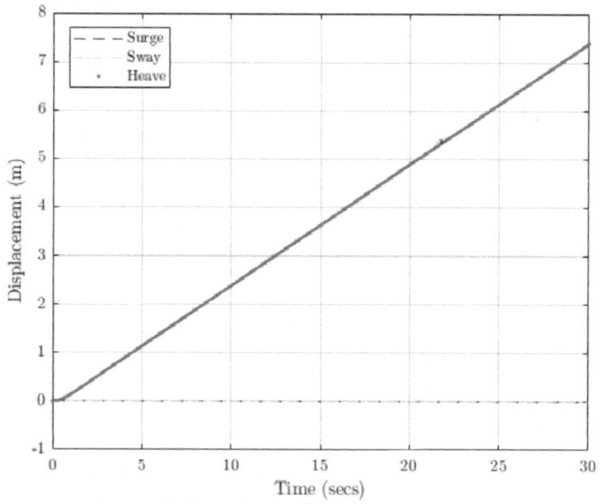

Figure 3: Test 02 – displacement response

Test 03: Thrust force was applied by the axial thrusters on the four HAUVs such that Seaperch 01 and 03 applied their positive thrust whereas Seaperch 02 and 04 applied the negative thrust. The system moves in the axial as well as in the transverse direction due to the yaw angle produced, as shown in Figure 4 and Figure 5 respectively. The trajectory made by the system in the xy-plane is shown in Figure 6.

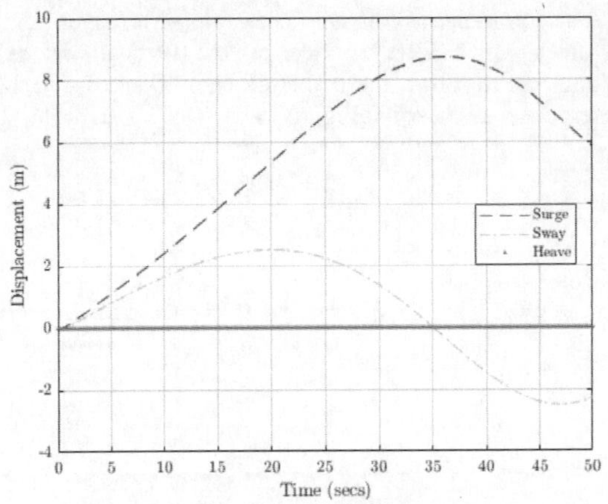

Figure 4: Test 03 – displacement response

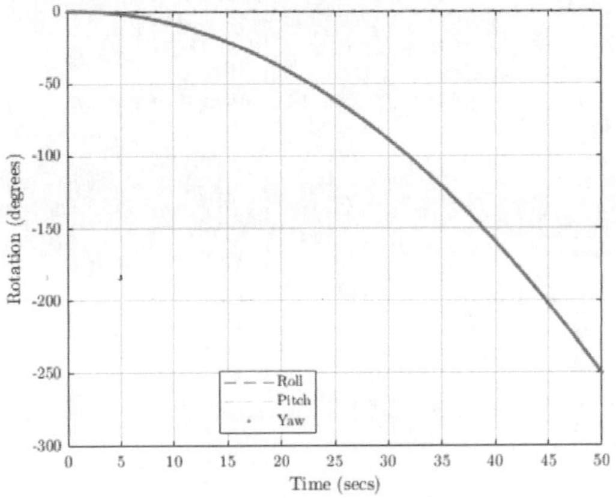

Figure 5: Test 03 – rotation response

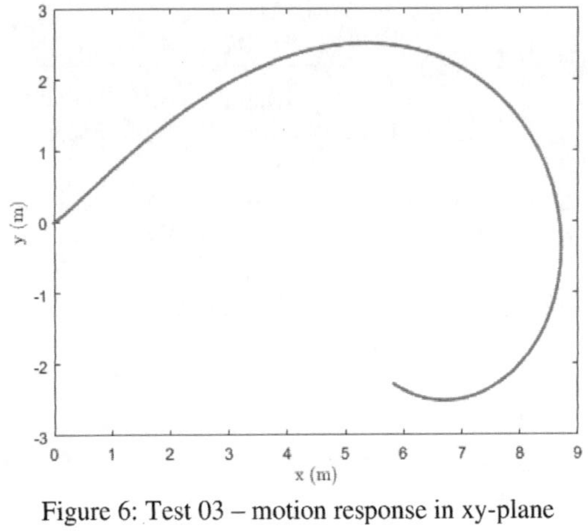

Figure 6: Test 03 – motion response in xy-plane

Test 01, 02 and 03 verify the motion response of the simulation model.

5. CONTROL SYSTEM DESIGN

To achieve the desired motion response, a control system was designed with the PID controllers to get the desired state response. The control strategy is shown in Figure 7.

- The PID controllers are applied on the error ($\boldsymbol{e}$) which is the difference between the desired and actual state of the system. The outcome of the controllers is given to $\boldsymbol{\tau_e}$ which is multiplied by the inverse of the transformation matrix ($\boldsymbol{J^{-1}}$) to get $\boldsymbol{\tau_d}$.
- $\boldsymbol{\tau_d}$ is multiplied by the inverse of thrust allocation matrix ($\boldsymbol{T_a}^{-1}$) to get $\boldsymbol{f_d}$. The saturation limits are applied on $\boldsymbol{f_d}$ to achieve $\boldsymbol{f}$ which is given to the propulsion model of the combined system to achieve $\boldsymbol{\tau}$.
- $\boldsymbol{\tau}$ is then given to the dynamic model to get $\boldsymbol{\eta}$ which is then compared with $\boldsymbol{\eta_d}$ to get $\boldsymbol{e}$.
- The process continues until the error gets to zero.

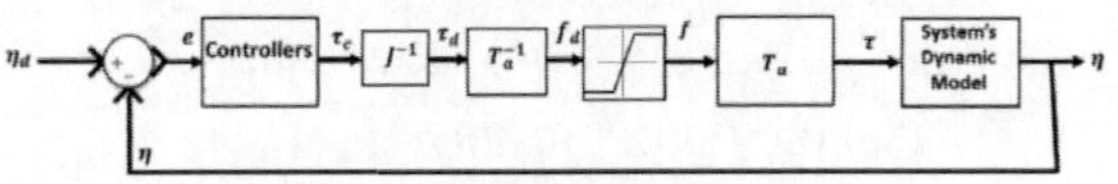

Figure 7: Control System

The PID controllers were applied in surge, sway, heave and yaw but not in roll and pitch due to the higher static stability of the system.

Test 04: In this test, the control system was applied to achieve the desired heave distance of 2m with zero motion in other directions. Figure 8 shows that the heave distance is achieved in around 12secs with the PID gains as mentioned in Table 2.

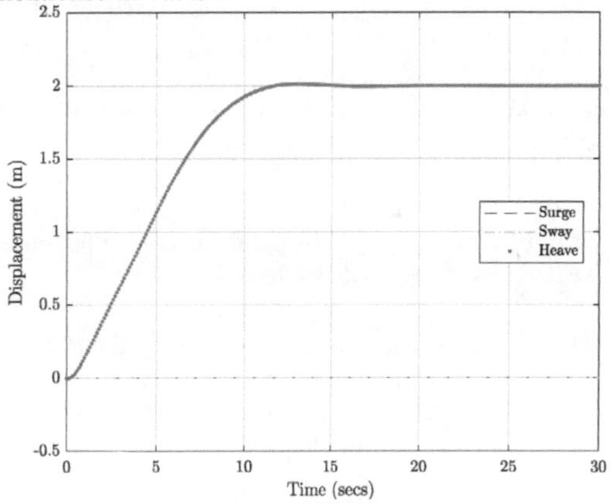

Figure 8: Test 04 - displacement response

Table 2: Test 04 - PID gains

	K_p	K_i	K_d
Surge	3	0	1
Sway	3	0	1
Heave	10	0	10
Yaw	3	0	1

From Figure 9, each vertical thrusters initially apply at the maximum thrust force of 0.8N which makes a total of 6.4N. This thrust force is applied for a duration of 5secs. Then, it gradually starts decreasing as the system approaches the desired vertical distance. A small negative thrust is experienced between 10secs and 14secs to stop the system and keep it at the desired location.

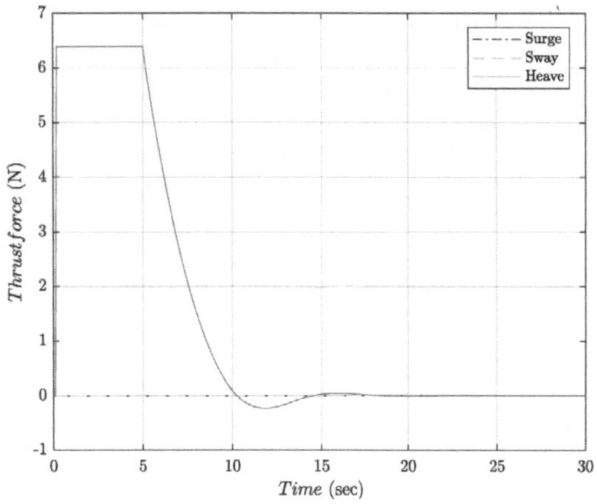

Figure 9: Test 04 - force response

6. TRAJECTORY GENERATION

In the above analysis, the control system was given the task to reach the desired position. However, the path followed by the system was not controlled which could have the obstacles. Therefore, it is required to develop a safe trajectory and controllers are applied to follow it.

A minimum snap trajectory was generated in which the path is divided into segments. The path between the two waypoints in a segment is represented by a 7th order polynomial [10]. The duration of each segment was kept equal to 20secs and the time step was 0.01sec which is equal to the time step of the simulation.

Test 05: The control system was designed with the PID controllers to follow the trajectory. The whole path was divided into three segments. The first segment to move the system to a depth of 2m, the second to the axial distance of 4m and the third to bring it back to zero depth while keeping the same axial position.

Though the PID tuning methods are available to get a better choice of PID gains, however, the results in this paper were obtained by entering the PID gains manually. The reason is to observe the effect of increasing or decreasing PID gains. Initially, low PID gains were selected as shown in Table 3.

Table 3: Test 05 – low PID gains

	K_p	K_i	K_d
Surge	3	0	3
Sway	1	0	0
Heave	3	0	3
Yaw	1	0	0

The trajectory is followed with a time lag, as shown in Figure 10. During the first segment, the system takes around 27secs to reach the desired depth of 2m. During the second segment, the system reaches the desired surge distance of 4m in 48secs. Whereas the third segment which brings the system back to the 0m heave position is not followed in the allocated time. Therefore, the transportation task is unsuccessful. Moreover, thrust forces which were applied by the vertical and axial thrusters to follow the trajectory show slackness in their response, as shown in Figure 11.

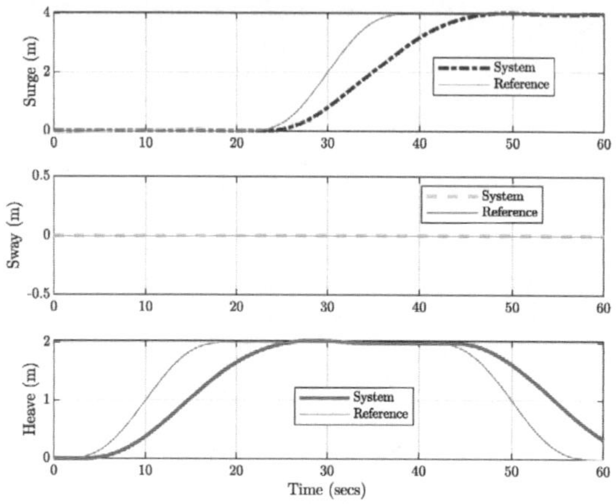

Figure 10: Test 05 – motion response

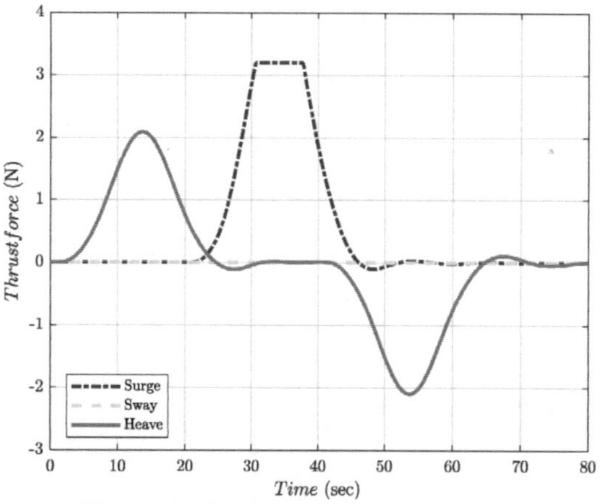

Figure 11: Test 05 – thrust force response

Test 06: Due to the limitations of test 05, the PID gains were increased for surge and heave, as shown in Table 4.

Table 4: Test 06 – higher PID gains

	K_p	K_i	K_d
Surge	10	0	10
Sway	1	0	0
Heave	50	0	40
Yaw	1	0	0

Figure 12 shows that the time lag while following the desired trajectory is reduced to a reasonable range. Therefore, the transportation task completes in time with the modified PID gains. Moreover, the thrust force slackness is reduced, as shown in Figure 13.

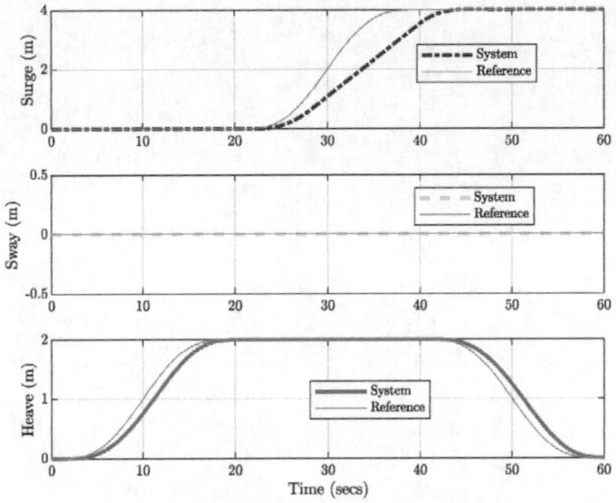

Figure 12: Test 06 – motion response

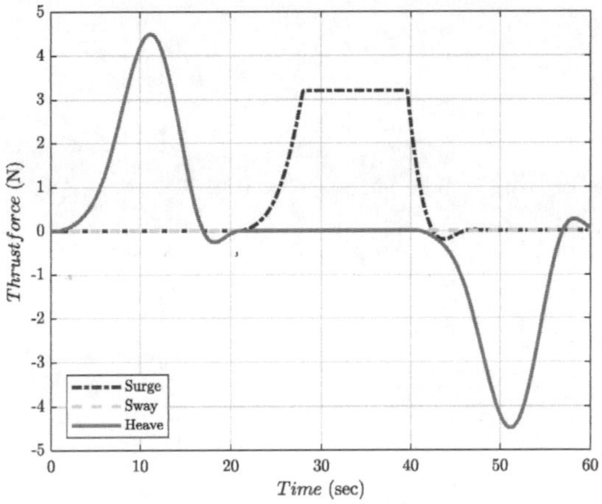

Figure 13: Test 06 – thrust force response

7. CONCLUSIONS

This paper developed a simulation model and designed a control system with PID controllers to control the motion response of four Seaperch HAUVs rigidly connected to a payload via solid plate links. This setup was selected to resemble the system which is being fabricated for experimental validation. The simulation model was first verified without implementing the control system. The control system was then designed with the PID controllers, which was proved to be robust enough to move the system to the desired location. However, when the system was commanded to follow a trajectory, the transportation task failed to complete in the allocated time at low PID gains due to the higher time lag. Therefore, the PID gains were increased to bring the time lag to a reasonable range and the transportation task was successfully completed. In this paper, the sea current and other possible disturbances were ignored which could be challenging for the controllers in the control system design. Therefore, it is recommended to include the disturbances in future work.

8. REFERENCES

[1] W. H. Wang, X. Q. Chen, A. Marburg, J. G. Chase, and C. E. Hann, "Design of Low-Cost Unmanned Underwater Vehicle for Shallow Waters," *IEEE/ASME Int. Conf. Mechatron. Embed. Syst. Appl.*, pp. 204–209, 2008.

[2] O. Yildiz, R. B. B. Gokalp, and A. E. E. Yilmaz, "A review on motion control of the Underwater Vehicles," *Electr. Electron. Eng.*, pp. II-337-II-341, 2009.

[3] J. Ghommam, H. Mehrjerdi, M. Saad, and F. Mnif, "Formation path following control of unicycle-type mobile robots," *Rob. Auton. Syst.*, vol. 58, no. 5, pp. 727–736, 2010.

[4] F. U. Rehman, G. Thomas, and E. Anderlini, "Development of a Simulation Platform for Underwater Transportation using two Hovering Autonomous Underwater Vehicles (HAUVs)," in *Proceedings of the 6th International Conference of Control, Dynamic Systems, and Robotics (CDSR'19)*, 2019, pp. 1–8.

[5] A. Phillips, M. Furlong, and S. R. Turnock, "The use of Computational Fluid Dynamics to Determine the Dynamic Stability of an Autonomous Underwater Vehicle," *10th Numer. Towing Tank Symp. (NuTTS'07), Hamburg, Ger.*, 2007.

[6] C. D. Williams, "AUV systems research at the NRC-IOT: An update," *Int. Symp. Underw. Technol.*, pp. 59–73, 2004.

[7] D. Mellinger, "Trajectory Generation and Control for Quadrotors," Mechanical Engineering and Applied Mechanics PhD Thesis, University of Pennsylvania, 2012.

[8] N. Michael, J. Fink, and V. Kumar, "Cooperative manipulation and transportation with aerial robots," *Auton. Robots*, pp. 1–14, 2010.

[9] T. I. Fossen, *Handbook of Marine Craft Hydrodynamics and Motion Control*, 1st Ed. Sussex: John Wiley & sons, 2011.

[10] F. U. Rehman, G. Thomas, and E. Anderlini, "Centralized Control System Design for Underwater Transportation using two Hovering Autonomous Underwater Vehicles (HAUVs)," *IFAC-PapersOnLine*, vol. 52, no. 11, pp. 13–18, 2019.

[11] J. Peraire and J. Widnall, "Lecture L26 - 3D Rigid Body Dynamics : The Inertia Tensor," in *Dynamics*, 2008.

[12] J. R. Morison, M. P. O'Brien, J. W. Johnson, and S. A. Schaf, "The force exerted by surface waves on piles," *Pet. Technol.*, vol. 2, no. 5, 1950.

[13] O. A. Eidsvik, "Identification of Hydrodynamic Parameters for Remotely Operated Vehicles," Marine Technology Master Thesis, Norwegian University of Science and Technology, 2015.

[14] S. S. Sandøy, "System Identification and State Estimation for ROV uDrone," Marine Technology Master Thesis, Norwegian University of Science and Technology, 2016.

9. AUTHORS BIOGRAPHY

Faheem Ur Rehman is currently undertaking PhD studies at UCL Mechanical Engineering department. His field of research is autonomous underwater transportation using multiple autonomous vehicles. He is a Member of the Royal Institute of Naval Architects (RINA) and a student member of the Society of Naval Architects and Marine Engineers (SNAME).

Dr Enrico Anderlini is a senior research associate at UCL Mechanical Engineering. He teaches ocean engineering at undergraduate and postgraduate level and provides support to the ship and submarine design exercises. During his doctorate, Enrico developed control strategies for wave energy converters using reinforcement learning. He is an Associate Member of the Royal Institute of Naval Architects (RINA).

Professor Giles Thomas is BMT Chair of Maritime Engineering in University College London's Department of Mechanical Engineering. A naval architect, his research focusses on the performance of ships, boats and offshore structures. Specific interests include fluid-structure interaction, hydrodynamics, full-scale measurements, model testing and design. He is a Fellow of the Royal Institution of Naval Architects (RINA) and a Chartered Engineer (CEng – Engineering Council of Great Britain).

Dr Saeed Khalid is a professor of Naval Architecture in Pakistan Navy Engineering College of National University of Science and Technology. His research focuses on marine structures, ship dynamics and submarine manoeuvring.

Dr Zeeshan Riaz is working as Dean of Naval Architecture department at Pakistan Navy Engineering College. His areas of interest are Computational Fluid Dynamics, UUV designing and submarine hydrodynamics.

NUMERICAL AND EXPERIMENTAL ANALYSES OF VARIABLE BUOYANCY SYSTEM FOR AUTONOMOUS UNDERWATER VEHICLE

B. K. Tiwari and **R. Sharma**, Indian Institute of Technology Madras, India

SUMMARY

Autonomous Underwater Vehicles/ Gliders (AUVs/Gs) are applicable for the underwater survey both in coastal areas and deep sea without any human intervention during their operations. However, their high energy consumptions required by the AUVs' thrusters for their maneuverability adversely affect their performance and endurance, etc. Herein, we present numerical and experimental results towards the design and development of a novel variable buoyancy system for autonomous underwater vehicle. Our design idea is based upon the 'Pump Driven Variable Buoyancy System (PDVBS)' and our results show that it is an energy efficient system offering excellent maneuverability. We focus especially on the vertical, forward and heave motions of the vehicle. Detailed performance analysis is presented along with the integration of the PDVBS with the AUV's design. Results are presented for the design of 5 kg buoyancy $(\pm B)$ capacity with a diaphragm type PDP at the maximum buoyancy change rate of 5.0 kg/min and the depth control performance of the AUV to hover at desired depth of 60 m using the LQR controller are analyzed in-detail.

NOMENCLATURE

$\pm B$ = Total changeable buoyancy capacity (kg)
w = Heave velocity (m/s)
W = Weight of the AUV (N)
B = Buoyancy force acting (N)
U = Speed of the vehicle (m/s)
M = Pitching moment (N-m)
Z = Vertical forces (N)
L = Length of the AUV (m)
D_{max} = Maximum diameter of the vehicle (m)
C_d = Drag coefficient (non-dimension)
ρ = Density of the fluid (kg/m^3)
∇ = Displaced volume (m^3)
g = Gravitational acceleration (m/s^2)
b = Diameter of the vehicle along x-axis (m)
m = Mass of the vehicle (kg)
z_G = Center of gravity in z-direction (m)

1. INTRODUCTION

High energy consumptions are required by the AUV's propeller for their maneuverability performance during the operation. In this environment 'Variable Buoyancy System (VBS)' is expected to minimize the energy consumption during the AUV's operations such as sinking, rising or station keeping at any desired depth without operating the thrusters and for the saw-tooth motion of gliders.

Seahorse AUV [1] was developed by the ARL/USA, which was very large in size and designed for Naval Oceanographic Office (NAVOCEANO) with design requirements of low operating speed (i.e. 0.21 m/s). At low speed control surfaces are ineffective for trim control. Hence, it was integrated with very large buoyancy capacity of ($\pm B$ = 90 kg) VBS. Another very large AUV namely URASHIMA AUV was developed by the JAMSTEC for deep and long cruising range and it was integrated with buoyancy system of 50 liters capacity of oil transferable capacity to control the buoyancy of the vehicle. In this system oil is transferred from the oil tank to rubber bladder vice-versa to change buoyancy in order to compensate the effect of the variation in the seawater density. Later, the Theseus AUV was developed by International Submarine Engineering (ISE) and integrated with the buoyancy system of $\pm B$ = 95 kg, to compensate for the buoyancy change (i.e. decrease in weight) due to laying the underwater fiber optic cables.

However, all these buoyancy systems are very large and can be installed only in the very large size of the AUVs, e.g. beyond 8 m in length. Also some of the gliders such as Sea glider [3] and Spray gliders [4] are integrated with buoyancy system and control their buoyancy by changing the displaced volume of the external bladders in the order of around 200 ml.

Also, the displaced volume of metal bellows [5] can be changed to control the buoyancy either by operating electric linear actuator or by using paraffin wax with peltier device method. Slocum glider uses single-stroke displacement piston pump with rolling diaphragm seal to move water and control the buoyancy by changing the overall weight of the glider. However, we have noticed that that all of these are limited to very low capacity of buoyancy control in the area of gliders.

Herein, we focus on the design and development of the VBS of $\pm B$ = 5 kg. This can be integrated to the UVs and make them capable to perform the following tasks:

- Pitch control (i.e. trimming the vehicle, because at lower speed of the vehicle control surface becomes ineffective).
- Rate of change of depth in control (i.e. change in the heave speed, both in diving and climbing).

- Emergency release of recovery (by releasing the weight).
- Station keeping in the vertical water column at desired depths.
- Compensating for the changes in salinity and density.

Herein, a novel and simple technical approach is used to control the net change in buoyancy force by controlling the overall mass of the vehicle. Remaining part of the paper organized as: Section 2 discusses the methodology, mathematical modeling of the VBS in stand-alone mode and VBS model for the trim/depth control of operating AUV, Section 3 presents mathematical modeling of the AUV integrated with VBS, Section 4 analyzes the simulation results and discussion, Section 5 presents the path planning of the AUV, and Section 6 concludes the paper with some identifications about the future scope of research.

2. METHODOLOGY

From fundamental of physics [5], net buoyancy of a submerged body/vehicle is defined:

$$\Delta B = W - B = (m - \rho \times \nabla) \times g \quad (1)$$

where W is the total weight of vehicle, B is the buoyancy of the vehicle, ∇ is the volume displaced by the vehicle, g is the gravitational acceleration, and ρ is the density of the fluid. If the net buoyant force $\Delta B = 0$, then the vehicle is neutrally buoyant, if $\Delta B < 0$ then it is positively buoyant, and if $\Delta B > 0$ then the vehicle is negatively buoyant.

So, in order to control the depth, based upon the operational requirements we need to change ΔB and this can be done either by changing the mass of the vehicle m or by changing the volume displaced by the vehicle ∇. Different methods to control the ∇ are:

- Hydraulic driven VBS (oil tank and rubber bladder with hydraulic mechanism for fluid transfer in either way).
- Pneumatic driven VBS (expending/compressing the rubber bladder by high pressure compressed air). At every cycle the volume of the air exhausted mass of the air also changes in every cycle and pressure decreases of the high pressure chamber.
- Using the metal bellow (expanding/compressing the metal bellows by linear actuator or paraffin wax with peltier device) and
- Moving the hollow piston.

Another method of buoyancy control, i.e. by changing in m can be achieved by two methods:

- Releasing the dead weight (only one way buoyancy can be control and suitable for emergency case only).
- Water ballast tank fill/empty (easy to control the mass of the vehicle by filling/emptying the ballast tank, good for large buoyancy capacity).

This paper focuses on change of the mass of the vehicle and the proposed designed VBS consist of three modular parts: 1) upper enclosed dome, 2) middle part and 3) hemispherical ballast tank. Main components of the designed VBS includes semispherical ballast tank, diaphragm pump, 12-V DC solenoid valve to control the flow direction, flow sensor, check valves, micro controller and data logging system. CAD Model of the developed VBS of $\pm B$ = 5 kg buoyancy with its modular components is shown in Fig. 1(a) and the leakage and stability testing of the VBS in stand-alone mode is shown in Fig. 1 (b). Inner radius of the hemispherical ballast tank is 0.150 m and thickness is 0.006 m and the material used for the tank is e-FRP.

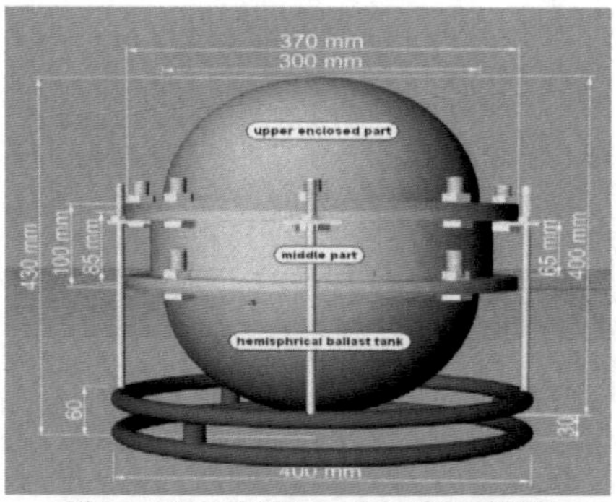

Figure 1(a): CAD model of the VBS.

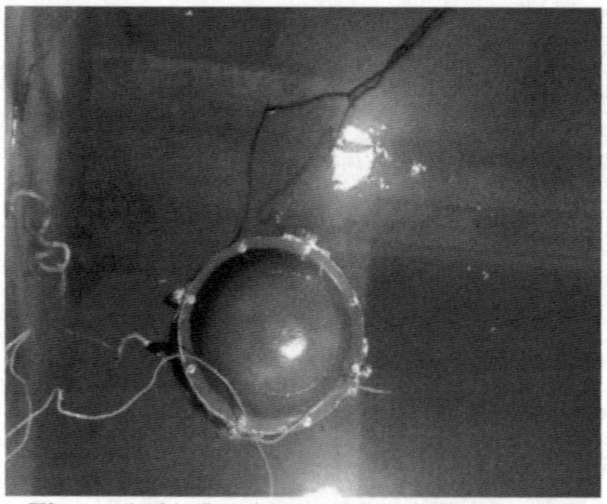

Figure 1 (b): Leakage and stability testing of the VBS in stand-alone mode.

2.1 MATHEMATICAL MODELING OF THE VBS IN STAND-ALONE MODE

Here, we develop the mathematical model for the VBS in stand-alone mode to study the detail analysis and performance of the VBS, and it is:

$$\begin{aligned}(m + \Delta m)\cdot a &= F_G - F_B - F_D - F_a \\ &= (m + \Delta m) g - \rho_{sw} V_{di} g \\ &\quad - \frac{1}{2} C_D \rho_{sw} A_p |w| w - m_a a\end{aligned} \quad (2)$$

where m is mass of the vessel, Δm is the change in mass, F_G is the gravitational force, F_B is the buoyancy force, F_D is the drag force, g is the gravitational acceleration, ρ_{sw} is the sea water density, V_{di} is the displaced volume by the vessel, C_D is the drag coefficient, w is the velocity in the vertical plane, m_a is the added mass and a is the acceleration of vehicle. Assuming $w=0$ at $\Delta m=0$ (i.e. neutrally buoyant and $mg=\rho_{sw}V_{di}g$), we get:

$$(m+\Delta m+m_a)\bullet a=\Delta mg-\frac{1}{2}C_D\rho_{sw}A_p|v|v \tag{3}$$

$$a=\frac{\left(-1/2\bullet C_D\rho_{sw}A_p|w|w\right)}{(m+\Delta m+m_a)}+\frac{\Delta mg}{(m+\Delta m+m_a)} \tag{4}$$

$$\dot{w}=-\frac{C_D\rho_{sw}A_p|w|w}{2\left(m+m_a+\frac{\delta W}{g}\right)}+\frac{\delta W}{\left(m+m_a+\frac{\delta W}{g}\right)} \tag{5}$$

$\dot{z}=-u\theta+w$, assuming $u=0$ i.e. vertical plane only, $\dot{z}=w$. (6)

Equation (5) and (6) can be written in the state space form as the following:

$$\begin{bmatrix}\dot{z}\\ \dot{w}\end{bmatrix}=\begin{bmatrix}0 & 1\\ 0 & -Z_v\end{bmatrix}\begin{bmatrix}z\\ w\end{bmatrix}+\begin{bmatrix}0\\ K_{in}\end{bmatrix}\delta W \tag{7}$$

where $Z_v=\dfrac{C_D\rho_{sw}A_p|v|}{2\left(m+m_a+\frac{\delta W}{g}\right)}$ & $K_{in}=\dfrac{1}{\left(m+m_a+\frac{\delta W}{g}\right)}$.

Simulation results for the VBS in stand-alone mode are shown in the Fig. 2. Here the heave/terminal velocity versus time for 1, 1.5 and 2 kg negatively buoyant VBS is shown in Fig 2 (a) and the maximum heave velocity achieved is around 0.8 m/s for 2 kg buoyancy and lowest is 0.55 m/s at 1 kg of negative buoyancy.

Open loop simulation results for the depth versus time of VBS are shown in Fig. 2 (b). From these results we observe that the depth is linearly increasing with time.

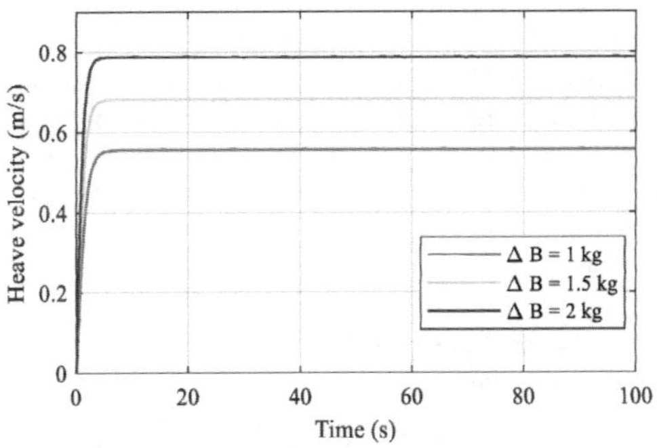

Figure 2 (a): Open loop simulation results of heave velocity versus time of VBS in stand-alone mode.

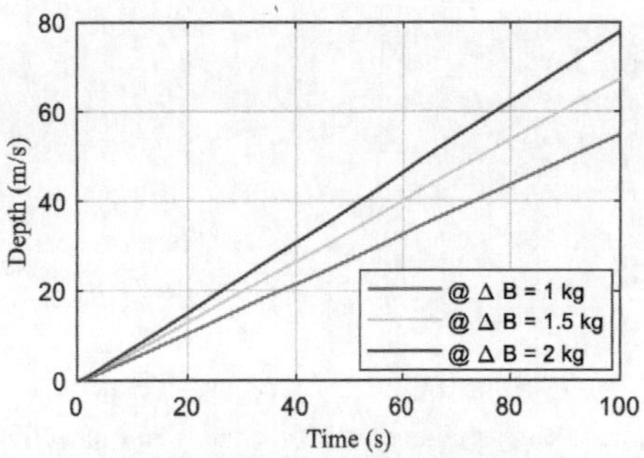

Figure 2 (b): Open loop simulation results for the depth versus time of VBS.

2.2 VBS MODEL FOR TRIM/DEPTH CONTROL OF OPERATING AUV

Buoyancy model as shown in Fig. 3 presents the force or moment generated by the VBS when it operates. In these operations the IMU sensor sends the signal for requirement of increase or decrease in the buoyancy in order to control the depth of operation or pitch. Furthermore, after considerable time delay of valve and operating the pump at any specified flow rate, integrating the flow rate for a specified time interval will result into the desired amount of buoyancy change. This implies into change of the buoyancy force and also the moments about the CG of the vehicle in order to control the vertical motion as well as the attitude or trimming of the vehicle.

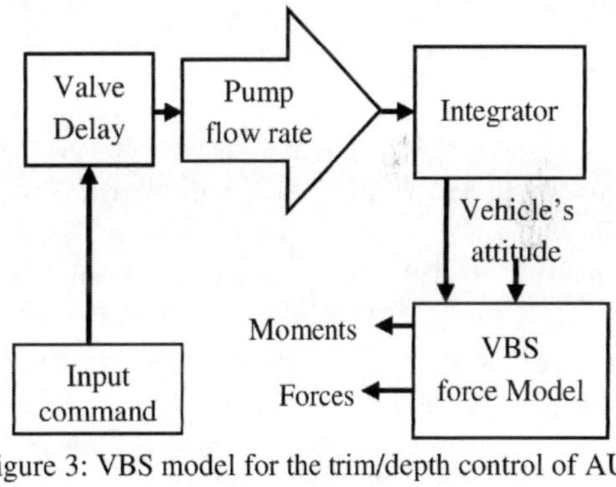

Figure 3: VBS model for the trim/depth control of AUV in operation.

3. MATHEMATICAL MODELING OF THE AUV INTEGRATED WITH VBS

Basic geometrical shape of AUV which is integrated with two VBS, one is at the nose side at distance of L_1 and the other is at distance L_2 from the CG of the vehicle is shown in Fig 4. Buoyancy capacity of each VBS integrated is 5 kg only and the detail analysis of AUV is performed to achieve the desired depth of 60 m and hover at the same without operating the propeller. Geometric details are the following:

$$r_n = \frac{D_{max}}{2}\left\{1-\left(\frac{L_n - x}{L_n}\right)^{n_n}\right\}^{1/n_n} \quad ; for\ 0 \le x \le L_n, \tag{8}$$

$$r_m = \frac{D_{max}}{2}; for\ L_n \le x \le (L_n + L_m), \tag{9}$$

$$r_t = \frac{D_{max}}{2}\left\{1-\left(\frac{x - L_n - L_m}{L_t}\right)^{n_t}\right\}; (L_n + L_m) \le x \le L \tag{10}$$

where r_n, r_t, n_n, n_t D_{max}, L_n, L_m, L_t are the nose radius, tail radius, nose shape coefficient, tail shape coefficient, maximum diameter, nose length, tail length and middle length of the AUV respectively. We consider length of the vehicle 3.0 m to study the simulation performance of the AUV integrated with VBS. Mathematical modeling of this is divided into two parts, i.e. kinematics and dynamics of the AUV.

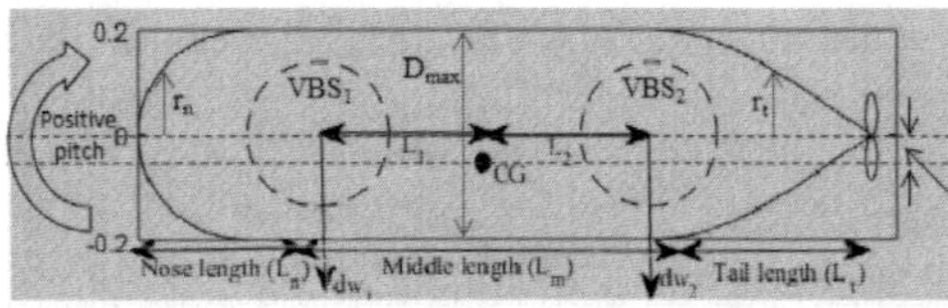

Figure 4: Basic geometrical shape of the AUV integrated with VBS.

3.1 KINEMATIC EQUATION

Kinematic equation of motion in the six DOF [6] of an underwater vehicle can be in the body fixed frame and this is:

$$\begin{pmatrix}\dot{\eta}_1 \\ \dot{\eta}_2\end{pmatrix} = \begin{pmatrix} J_1(\eta_2) & 0 \\ 0 & J_2(\eta_2)\end{pmatrix}\begin{pmatrix} v_1 \\ v_2 \end{pmatrix} \tag{11}$$

where $\eta_1 = [x, y, z]$ and $\eta_2 = [\phi, \theta, \psi]$ are the linear and angular position vectors in earth fixed frame, $v_1 = [u, v, w]$ and $v_2 = [p, q, r]$ are the linear and angular velocity vectors in the body fixed frame, respectively. The $J_1(\eta_2)$, $J_2(\eta_2)$ are the rotational (transformation) matrices in which $J_1(\eta_2)$ is skew-symmetric matrix. For analyzing the performance of the AUV integrated with VBS, we note that the buoyancy system influences the motion of the vehicle in vertical plane only. Hence, the kinematic equation of the AUV is taken in the vertical plane only and that is:

$$\dot{\theta} = q, \tag{12}$$

$$\dot{z} = -u\, sin\theta + w\, cos\theta. \tag{13}$$

3.2 DYANMIC EQUATION

Dynamics equation for the force in Z direction and moment of the vehicle about the Y axis can be written:

$$Z = m(\dot{w} - uq + vr - z_G\left(p^2 + q^2\right) + x_G\left(pr - \dot{q}\right) + y_G(qr + \dot{p})) \quad \text{and} \tag{14}$$

$$M = I_y\dot{q} - (I_z - I_x)rp + I_{yz}(pq - \dot{r}) - I_{zx}\left(r^2 - p^2\right) + m\left[z_G\left(\dot{u} - vr + wq\right) - x_G\left(\dot{w} + vp - uq\right)\right]. \tag{15}$$

Restricting the motion in the vertical plane only (i.e. sway velocity $v = 0$, yaw rate $r = 0$, $\dot{u} = 0$ and $x_G = 0$) and neglecting the non-linear terms we get:

$$Z = m(\dot{w} - uq - x_G\dot{q}) \text{ and } M = I_y\dot{q} - mx_G(\dot{w} - uq) \tag{16}$$

Summation of the external forced acting on the vehicle consists of hydrodynamic force, drag force, gravity force, propeller force and force due to rudder deflection. However, in this study we neglect the propeller force and assuming that the vehicle is at very low speed the rudder also becomes ineffective. Now, the remaining external forces can be written as:

$$\sum Z = Z_q{}''Uq + Z_{\dot{q}}{}''\dot{q} + Z_w{}''Uw + Z_{\dot{w}}{}''\,\dot{w} + \underbrace{(W - B)\cos\theta}_{gravity\ component} \underbrace{-0.5\rho\int_{nose}^{tail} C_D b(x)(w - xq)\left|w - xq\right| dx}_{drag\ component} + \underbrace{(dw_1 + dw_2)}_{\Delta B\ component} \tag{17}$$

where $\sum Z$ is the sum of all the external forces acting on the vehicle in vertical plane, $Z_q{}''$, $Z_{\dot{q}}{}''$, $Z_w{}''$, $Z_{\dot{w}}{}''$, are the non-dimensional hydrodynamic coefficients, $\dot{q}, \dot{w}$ are the rate of change of pitch and heave velocities, respectively, C_D is the drag coefficient, $b(x)$ is the diameter of the vehicle along the x-axis, and dw_1, dw_2 are the weight of the added to ballast tank 1 and 2 respectively.

Moment acting in vertical plane due to external forces is:

$$\sum M = 0.5\rho\int_{nose}^{tail} C_D b(x)(w - xq)\left|w - xq\right| x\, dx + M_q{}'Uq + M_{\dot{q}}{}'\,\dot{q} + M_w{}'\,Uw + M_{\dot{w}}{}'\,\dot{w} + (x_G W - x_B B)\cos\theta - (z_G W - z_B B)\sin\theta + \underbrace{(-L_1 \times dw_1 + L_2 \times dw_2)}_{due\ to\ weight\ addition} \tag{18}$$

where L_1, L_2 are the distances of the CG of the VBS_1 and VBS_2 from CG of the vehicle respectively, and x_G, x_B, z_G, z_B are the distances between the CG and CB of the vehicle along the x and z axis from origin of the vehicle respectively. Furthermore, $M_{\dot{q}}{}'$, $M_w{}'$, $M_{\dot{w}}{}'$ are the non-dimensional hydrodynamic parameters acting during the pitching moment of the vehicle and other parameters are same as defined before. These parameters can be computed either experimentally (i.e. PMM test or Maneuvering tests) or numerically (i.e. using the CFD or strip theory etc.).

Herein, we use Beyazay [7], and the non-dimensional hydrodynamic parameters used in our simulation study are:

$$Z_{\dot{q}}' = -0.00253, Z_{\dot{w}}' = -0.0934, Z_{q}' = -0.0701,$$

$$Z_{w}' = -0.7844, M_{\dot{q}}' = -0.00625, M_{\dot{w}}' = Z_{\dot{q}}',$$

$$M_{q}' = -0.0153 \text{ and } M_{w}' = 0.0512$$

Simulation study is performed for the developed VBS of comparatively large buoyancy capacity which is integrated with 3.0 m length of the AUV. This will result into large heave velocity, increase the payload capacity and reduce the energy consumption comparatively to the thruster driven AUV.

3.3 CONTROLLER DESIGN

We use the MIMO optimal controller for the linearized system (i.e. for the linearized form of the AUV integrated with VBS) to perform computer simulation study in the vertical plane only. The performance index of the optical controller [8] for the linear continuous system

$$\dot{x} = A\bar{x} + B\bar{u} \tag{19}$$

where A is the state matrix, $\bar{x}$ is the state vector, B is the input matrix and $\bar{u}$ is the input vector, which is defined in the time domain of $t \in [t_0, t_1]$ and written as:

$$J = \int_0^\infty \left(\bar{x}^T Q \bar{x} + \bar{u}^T R \bar{u}\right) dt \tag{20}$$

where Q is the state weighing matrix and R is the control input weighting matrix. The Q and R are the relative importance parameters instead of their absolute values. Control input to system can be written as:

$$\bar{u} = -K_c \times \bar{x}_{error} \tag{21}$$

where K_c is the optimal controller gain matrix and $\bar{x}_{error} = \bar{x}_d - \bar{x}_{act}$ where K_c is computed as $K_c = R^{-1}B^T S \bar{x}$ in which S is associated with the Algebraic Riccati equation and this can be written as:

$$A^T S + SA - SBR^{-1}B^T S + Q = 0\,. \tag{22}$$

4. SIMULATION RESULTS AND DISCUSSION

4.1 OPEN LOOP OPERATION

Herein, we study the VBS integrated with AUV of 3.0 m length and results are presented in the open loop (i.e. zero reference in all the simulation results means that the vehicle is neutrally buoyant - half-filled ballast tanks. Fig. 5 (a) shows the weight addition to the ballast tanks, i.e. 1.5, 2.0 and 2.5 kg at the rate of 5 kg/min. This is clearly observed from Fig. 5 (b) that 2.5 kg mass is added to the tank in 30 sec and 1.5 kg in 18 s. However, in open loop operation once the ballast tanks are completely filled, vehicle starts sinking at constant/terminal velocity as shown in Fig. 6 (a). In Fig. 6 (a) for all the three cases of buoyancy addition, the results are reported for the sinking speed of 0.2, 0.23 and 0.27 m/s respectively and depth is observed to be continuously increasing with respective terminal velocity.

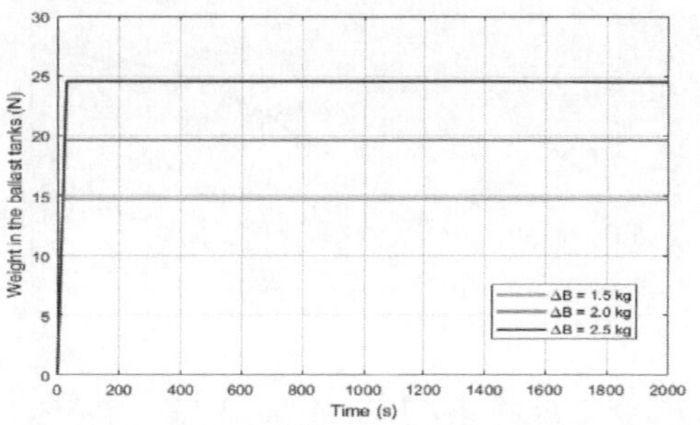

Figure 5(a): Variation of the weight addition in ballast tank versus time in open loop operation.

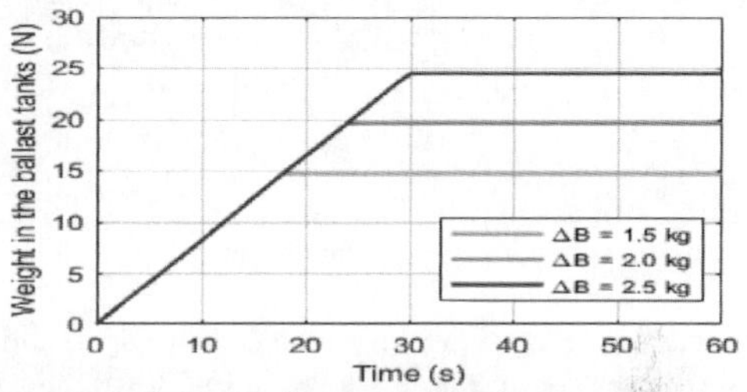

Figure 5 (b): Zoomed part of the weight addition in ballast tank versus time.

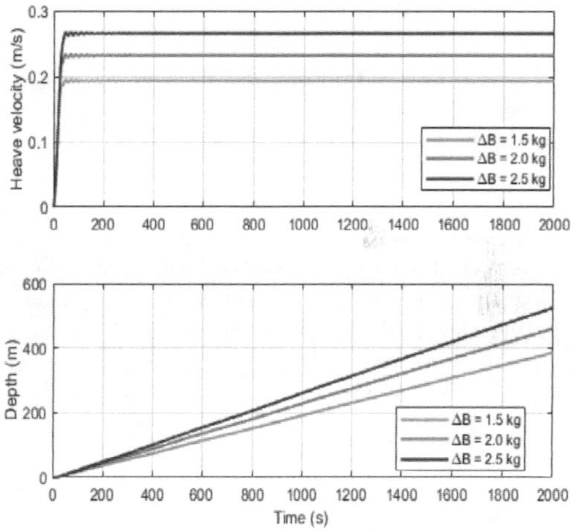

Figure 6 (a): Variation of the heave velocity and depth of operation versus time in open loop.

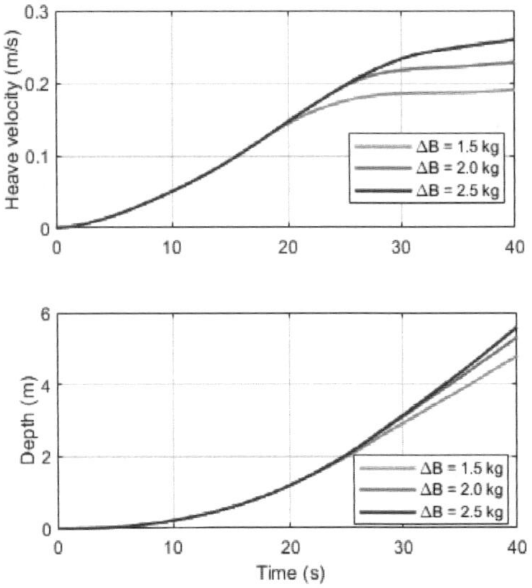

Figure 6 (b): Zoom part of the heave velocity and depth of operation versus time in open loop.

Zoomed part of the heave velocity and depth of operation versus time in open loop is shown in Fig. 6 (b), and we observe that until the ballast tanks are completely filled and heave velocity reaches to its maximum, depth is non-linearly varying with time and after that depth is linearly changing with time.

4.2 CLOSED LOOP OPERATION

Although, the VBS developed in this study is capable of operating at higher depths also, because the selected positive displacement pump used is limited to operate only up to 6.5 bar (i.e. 65 m depth) pressure, the performance of the AUV integrated with the designed VBS at different buoyancy capacities is analyzed to control the depth of operation at 60 m. Fig. 8 shows the weight addition to the ballast tanks, from which we can observe that the weight added to ballast tank for three different capacity 1.5, 2.0 and 2.5 kg, remains constant and the AUV starts sinking at the terminal velocity as shown in Fig 7. As the vehicle is approaching the desired depth operation, weight of the water inside the ballast tank starts reducing and finally reached to zero (i.e. half-filled and neutral condition). From Fig. 7 we also observe that the vehicle reaches to the desired depth and gets an overshoot of 1.5 m. Then the controller performs the action to bring it to the desired depth and finally this gets settled at 60 m after 800 s of sinking. From these results, we can observe that controller performance is better so that the desired depth of operation is achieved with almost zero overshoot.

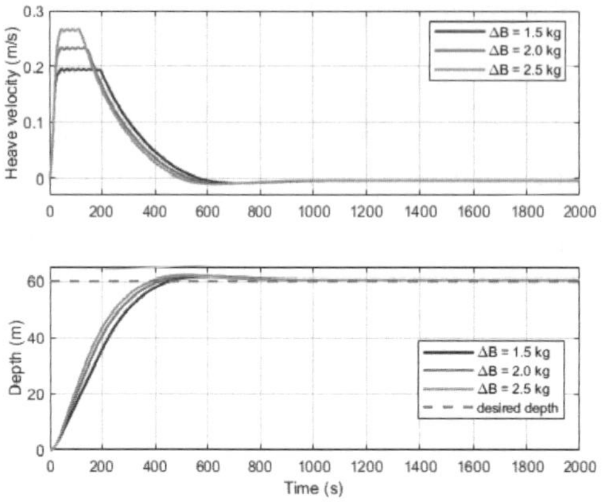

Figure 7: Variation of the heave velocity and depth of operation versus time in closed loop.

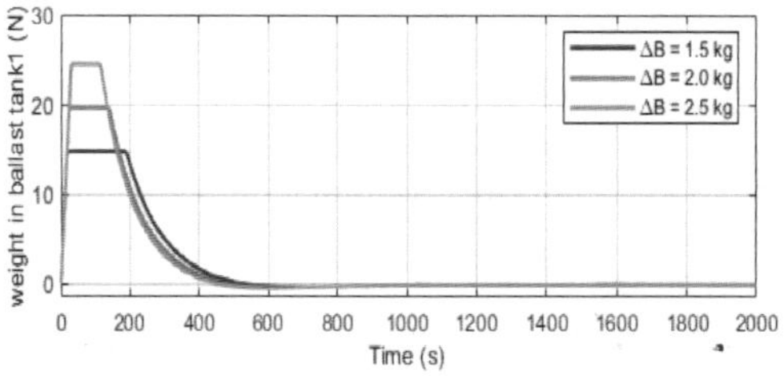

Figure 8: Variation of the weight addition in ballast tank versus time in close loop operation.

5. PATH PLANNING OF THE AUV

Path planning of robots is an important issue to be addressed for collision free motion in the environment and that can consist of different obstacles [9-10]. Herein, the planning analysis of the robot, especially AUVs, is studied to accomplish the obstacles avoidance for desired motion the vehicle. Although the AUV consists of various sensors such as SONAR (i.e. for detecting the obstacle, underwater communication and navigate the underwater), but for collision free motions of the underwater vehicles simplest search algorithm method called as classical graph-search algorithms (also known as grid-based search algorithm) are preferred. Classical graph-search algorithms are further classified as: 1) Uninformed search algorithms which include Depth-First Search (DFS), and Breadth-First Search (BFS), etc.

Depth-First Search (DFS) algorithm is very similar to BF search but only the difference is in the choice of the next point to compute. In depth-first algorithm the priority of the point is given heuristically by a distance function which is an estimate of the residual distance between the source point and the goal point. This method focuses more towards the directly search the target point instead of developing a front around the start point. It is not very optimal because reach the solution number of steps are high hence this is not optimal. Breadth-First Search (BFS) also called as Dijkstra's single source shortest-path algorithm which explores the next grid points for their distance from the source point at each step and next point is computed as the point whose distance is the lowest from the source and it is optimal if the costs of all edges are same. Informed search algorithms in which one have the information (i.e. heuristic) about the target state, this helps in more efficient searching. Heuristic cost is chosen based on the Euclidean distance between the point and the targeted goal divided by the maximum possible nominal speed which always results into the lower than the actual cost. The A^* algorithm is one of the most popular informed search algorithms used in the artificial intelligence for path planning. This is producing more efficient results for path planning of the AUVs.

We note here that the VBS presented in this work is integrable with the design and dynamics of AUV and also discussed the usefulness of the path planning for collision avoidance. It is concluded that the designed AUV need to be integrated with A^* algorithm and at present we are working on it.

6. CONCLUSION

We presented the design and development of the VBS for the buoyancy capacity of $\pm B = 5$ kg. Simulation performance analyses for both the designed VBS in stand-alone mode and the AUV integrated with the VBS at different buoyancy capacities were shown to offer excellent performances. Depth of the AUV was controlled at 60 m desired depth using the state feedback, i.e. LQR controller. Although, the simulation study showed the good performance of the AUV and achieved the desired depth with almost zero overshoot, these are to be confirmed in-detail with experiments. Experimental testing of the presented design, integration with A^* algorithm, and analysis of the natural system behavior and coupling effects are some of our future areas of focus. Some of these are currently being investigated by us.

7. ACKNOWLEDGEMENTS

This research was supported by the internal research grants of IIT madras through research scheme: OE14D212 and from marine Systems Panel, NRB, India via a sponsored project: NRB-263/MAR/12-13

8. AUTHORS BIOGRAPHY

B.K. Tiwari received his BSc (Physics, Chemistry and Mathematics) from the Avadh University, Faizabad, India, in 2007 and BE (Aeronautical engineering) from the Aeronautical Society of India, in 2013. Currently, he is pursuing his PhD at the Department of Ocean Engineering, Indian Institute of Technology Madras, Chennai, India. His research interests are: Design and development of autonomous underwater vehicles and gliders, guidance and control of marine vehicles, marine robotics and mathematical modeling and simulation.

R. Sharma received his BE (Civil Engineering) from the Indian Institute of Technology, Roorkee (formerly University of Roorkee) in 1994; M.Tech (Ocean Engineering) in 1999 and PhD from the Indian Institute of Technology Kharagpur, India, in 2008. Presently, he is working as Professor at the Department of Ocean Engineering, Indian Institute of Technology madras, Chennai, India. His research interests are: Design, analysis and production of marine structures, marine robotics and dynamic data driven forecasting systems.

REFERENCES

[1] DZIELSKI, E., TANGIRALA, C., MOYER, W. W., and BRADLEY, D. L., 'NAVOCEANO Seahorse AUV Design, Testing, and Capabilities', *OCEANS 2002 MTS/IEEE Conference & Exhibition*, Vol. 1, pp. 151-155, 2002.

[2] HYAKUDOME, T., AOKI T., MURASHIMA, T., TSUKIOKA, S., YOSHIDA, H. and NAKAJOH, H., et al., 'Key Technologies for AUV URASHIMA', *Oceans Conference (IEEE),* Vol. 1, pp. 162–166, 2002.

[3] ERIKSEN, C.C., OSSE, T.J., LIGHT, T., WEN, R.D., LEHMANN, T.W., SABIN, P.L. and BALLARD J.W., 'Long Range Autonomous Underwater Vehicle for Oceanographic Research', *IEEE Journal of Oceanic Engineering*, Vol. 26 (4), pp. 424–436, 2001.

[4] DOUGLAS, C.W., SIMONETTI, P.J. and JONES, C.P., 'SLOCUM: An Underwater Glider Propelled by Environmental Energy', *IEEE Journal of Oceanic Engineering*, Vol. 26 (4), pp. 447–453. 2001.

[5] SHIBUYA, K., KISHIMOTO, Y. and YOSHII S. 'Depth Control of Underwater Robot with Metal Bellows Mechanism for Buoyancy Control Device Utilizing Phase Transition', *Journal of Robotics and Mechatronics*, Vol. 25(5), pp. 795-803, 2013.

[6] FOSSEN, T.I., Handbook of Marine Craft Hydrodynamics and Motion Control. Published by John Wiley & Sons, 2011

[7] BEYAZAY, B., Simulation and Modeling of a Soft Grounding System for an AUV, MS Thesis, Naval Postgraduate School, USA, 1999

[8]RICHARD, C.D. and ROBORT, H.B., 'Modern control systems 12th edition', published by Prentice Hall 2010.

[9] GARAU, B., ALVAREZ, A., and OLIVER, G, Path planning of Autonomous Underwater Vehicles in Current Fields with Complex Spatial Variability: An A^* Approach, Proceedings of *IEEE International Conference on Robotics and Automation,* pp. 194-198, 2005.

[10] VASUDEVAN, C., and GANESAN, K., Case-Based Path Planning For Autonomous Underwater Vehicles, *Autonomous Robots*, vol. 3(2) pp. 79-89, 1996.

THE IMPACT OF SEA CURRENT ON UNDERWATER TRANSPORTATION USING FOUR AUVS

F. U. Rehman, E. Anderlini and **G. Thomas,** University College London, UK

SUMMARY

This paper has investigated the impact of sea current on autonomous underwater transportation using four AUVs. A rigid connection method of multi-vehicular transportation was used. A nonlinear coupled dynamic model was developed to resemble the actual system. All the parameters were included by taking their effect about the centre of the combined body. Time-domain motion simulations were initially carried out without the sea current effect to verify the motion response and to work out the maximum payload mass which can be feasibly transported. The sea current effects were then applied in two conditions, firstly opposite to the direction of motion (condition 1) and secondly perpendicular to the direction of motion (condition 2). For condition 1, a limit was initially set to the payload mass before implementing the control system. A satisfactory motion response was obtained with the PID controllers. Whereas, the system response in condition 2 was not satisfactory due to the higher PID gains and undesirable pitch oscillations.

NOMENCLATURE

O	Centre of body
BFF	Body-fixed frame
EFF	Earth-fixed frame
COG	Centre of gravity
COB	Centre of buoyancy
$\boldsymbol{J}$	Complete transformation matrix
$\boldsymbol{M_{RB}}$	Rigid body mass matrix
$\boldsymbol{M_A}$	Added mass matrix
$\boldsymbol{M}$	Mass matrix
$\boldsymbol{C_{RB}(v_r)}$	Rigid body Coriolis matrix
$\boldsymbol{C_A(v_r)}$	Added mass Coriolis matrix
$\boldsymbol{C(v_r)}$	Combined Coriolis matrix
$\boldsymbol{D(v_r)}$	Damping matrix
$\boldsymbol{g(\eta)}$	Vector of hydrostatic parameters
$\boldsymbol{\tau}$	Thrust vector
$\boldsymbol{v}$	velocity vector
$\boldsymbol{v_r}$	relative velocity vector
m	mass of the system
W	weight of the system
B	Buoyancy of the system
(x_g, y_g, z_g)	Position of COG
(x_b, y_b, z_b)	Position of COB
(u_r, v_r, w_r)	Relative transitional velocities
(p_r, q_r, r_r)	Relative angular velocities
$(X_{\dot{u}}, Y_{\dot{v}}, Z_{\dot{w}})$	Translational added mass terms
$(K_{\dot{p}}, M_{\dot{q}}, N_{\dot{r}})$	Rotational added mass terms
$(X_{u\|u\|}, Y_{v\|v\|}, Z_{w\|w\|})$	Translational damping terms
$(K_{p\|p\|}, M_{q\|q\|}, N_{r\|r\|})$	Rotational damping terms
(τ_X, τ_Y, τ_Z)	Translational actuator terms
(τ_K, τ_M, τ_N)	Rotational actuator terms
(l_x, l_y, l_z)	Position of thruster w.r.t O
σ	Angle of axial thruster with horizontal
C_a	Added mass coefficient
C_D	Damping coefficient
$\boldsymbol{T_a}$	Combined thrust allocation matrix
$\boldsymbol{T_i}$ $i = 1,..4$	Thrust allocation matrix of each HAUV
K_P	Proportional gain
K_i	Integral gain
K_d	Differential gain
$\boldsymbol{\eta_d}$	Vector of desired state of the system
$\boldsymbol{\eta}$	Vector of actual state of the system
$\boldsymbol{e}$	Error between $\boldsymbol{\eta_d}$ and $\boldsymbol{\eta}$
$\boldsymbol{\tau_e}$	Thrust vector in EFF
$\boldsymbol{\tau_d}$	Thrust vector in BFF
$\boldsymbol{f_d}$	Desired thrust force vector in BFF
$\boldsymbol{f}$	Actual thrust force vector in BFF
$\boldsymbol{f_f}$	Vector of thrust force function

1. INTRODUCTION

The two types of Unmanned Underwater Vehicles (UUVs) are Remotely Operated Vehicles (ROVs) and Autonomous Underwater Vehicles (AUVs). ROV is a tethered, remotely controlled vehicle which is ideal for short-range precise operations [1]. Conversely, AUV is untethered, pre-programmed and suitable for long-range consistent operations [2]. ROV shaped vehicles can be used autonomously and are called Hovering Autonomous Underwater Vehicle (HAUV) [3] or Autonomous Underwater Helicopter (AUH) [4].

Payload transportation is an important requirement for various sectors, which is generally fulfilled by land vehicles, airplanes and sea surface vessels. Other than using giant vehicles, it is difficult to transport heavy payload by a single vehicle. However, with multiple agents, the heavier payload could be transported, as the load is distributed amongst the agents involved. The other advantages of using multiple vehicles are flexibility in using the number of vehicles and fault tolerance [5].

Two methods are mainly used for multi-vehicular transportation i.e. rigid body connection and flexible connection. In the rigid body connection method, all the vehicles are assumed to be rigidly connected to a central payload via solid links. In the development of the dynamic model, the whole system is assumed to be a single rigid body [6]. On the other hand, the flexible connection method assumes that the vehicles are connected to a solid payload by flexible cables. Each vehicle must operate independently to maintain its position and cable angle relative to the central payload [7].

The simulation model for multi-vehicular transportation in the underwater environment requires the calculation of the significant hydrodynamic and hydrostatic parameters. Moreover, the development of the mathematical model is challenging due to the nonlinear and coupled dynamics.

In this paper, Minerva ROV is assumed to be used autonomously and is called the Minerva HAUV (Hovering Autonomous Underwater Vehicle). This is a standard ROV operated by the Norwegian University of Science and Technology (NTNU) for which the model coefficients are known. The rigid connection method of multi-vehicular transportation is used such that four Minerva HAUVs are rigidly connected to a cubic payload via solid cylindrical links.

2. DEVELOPMENT OF DYNAMIC MODEL

A dynamic model was developed to analyse the motion response of the system of four Minerva HAUVs transporting a payload. The approach derived by Fossen [8] was adopted.

2.1. Reference frames

Two reference frames were defined, as shown in Figure 1. The position and orientation of the system were defined in EFF, whereas, the velocities, forces and their respective moments were expressed in BFF.

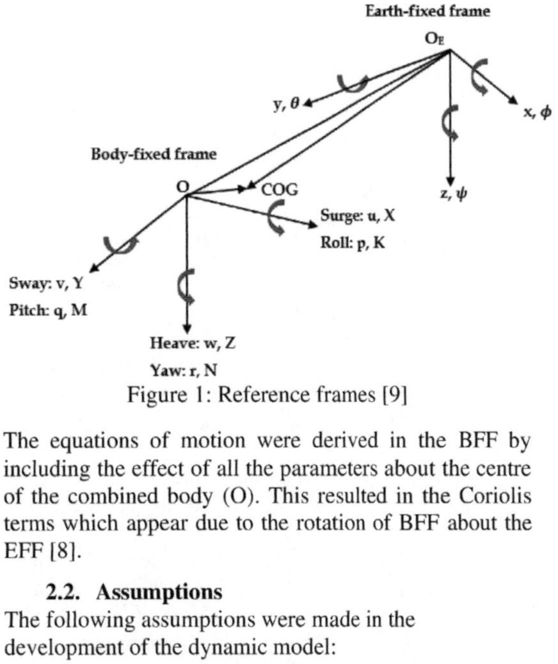

Figure 1: Reference frames [9]

The equations of motion were derived in the BFF by including the effect of all the parameters about the centre of the combined body (O). This resulted in the Coriolis terms which appear due to the rotation of BFF about the EFF [8].

2.2. Assumptions

The following assumptions were made in the development of the dynamic model:

1. The system is a rigid body of constant mass. This ensures that there is no relative motion between the mass particles [10]. Also, mass and its distribution do not change during the motion.
2. The vehicle is deeply submerged; therefore, the wave effects are ignored.
3. The interaction with other bodies is ignored.
4. The centre of the combined body is taken at the centre of the payload (O).
5. The linear damping terms are ignored.
6. Sea currents are applied at constant values.

The dynamic model was developed for the four Minerva HAUVs which are connected to a cubic payload via solid cylindrical links. The combined system has three planes of symmetry as the same HAUVs are attached through the same cylindrical links to a cubic payload in the centre, as shown in Figure 2. It was assumed that COG and O are at the same position, and COB is slightly above the COG to ensure static stability. Moreover, the combined product of inertia and cross-coupled hydrodynamic terms were negligible.

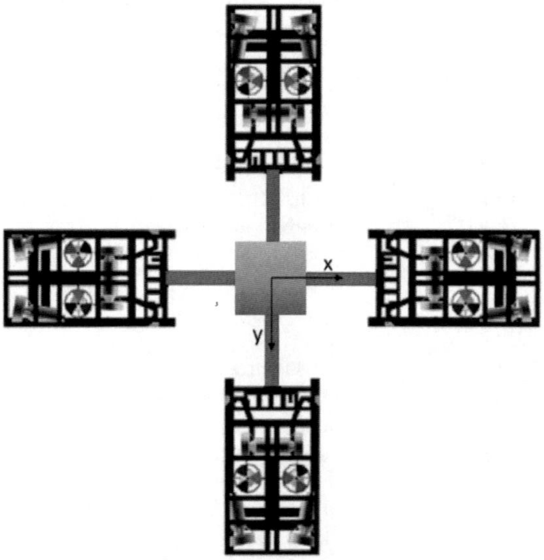

Figure 2: Four Minerva HAUVs connected to a cubic payload via cylindrical links

2.3. Equations of Motion

To understand different terms involved in the equations of motion and ease the development of the simulation model, the equations of motion are written in the matrices form as [8]

$$\boldsymbol{M}\dot{\boldsymbol{v}}_r + \boldsymbol{C}(\boldsymbol{v}_r)\boldsymbol{v}_r + \boldsymbol{D}(\boldsymbol{v}_r)\boldsymbol{v}_r + \boldsymbol{g}(\boldsymbol{\eta}) = \boldsymbol{\tau}. \qquad (1.1)$$

Where

$\boldsymbol{M} = \boldsymbol{M}_{RB} + \boldsymbol{M}_A.$

$\boldsymbol{C}(\boldsymbol{v}_r) = \boldsymbol{C}_{RB}(\boldsymbol{v}_r) + \boldsymbol{C}_A(\boldsymbol{v}_r).$

The relative velocity ($\boldsymbol{v}_r$) is the difference between the vehicle velocity ($\boldsymbol{v}$) and sea current velocity ($\boldsymbol{v}_c$) shown as

$$\boldsymbol{v}_r = \boldsymbol{v} - \boldsymbol{v}_c. \qquad (1.2)$$

Due to the three planes of symmetry, the matrices in equation (1.1) can be written in the reduced form as

$$\boldsymbol{M}_{RB} = [m, m, m, I_x, I_y, I_z]^{DT}. \qquad (1.3)$$

$$\boldsymbol{C_{RB}}(\boldsymbol{v_r}) = \begin{bmatrix} 0 & 0 & 0 & 0 & mw_r & -mv_r \\ 0 & 0 & 0 & -mw_r & 0 & mu_r \\ 0 & 0 & 0 & mv_r & -mu_r & 0 \\ 0 & mw_r & -mv_r & 0 & I_z r_r & -I_y q_r \\ -mw_r & 0 & mu_r & -I_z r_r & 0 & I_x p_r \\ mv_r & -mu_r & 0 & I_y q_r & -I_x p_r & 0 \end{bmatrix}. \quad (1.4)$$

$$\boldsymbol{M_A} = -[X_{\dot{u}}, Y_{\dot{v}}, Z_{\dot{w}}, K_{\dot{p}}, M_{\dot{q}}, N_{\dot{r}}]^{DT}. \quad (1.5)$$

$$\boldsymbol{C_A}(\boldsymbol{v_r}) = -\begin{bmatrix} 0 & 0 & 0 & 0 & Z_{\dot{w}}w_r & -Y_{\dot{v}}v_r \\ 0 & 0 & 0 & -Z_{\dot{w}}w_r & 0 & X_{\dot{u}}u_r \\ 0 & 0 & 0 & Y_{\dot{v}}v_r & -X_{\dot{u}}u_r & 0 \\ 0 & Z_{\dot{w}}w_r & -Y_{\dot{v}}v_r & 0 & N_{\dot{r}}r_r & -M_{\dot{q}}q_r \\ -Z_{\dot{w}}w_r & 0 & X_{\dot{u}}u_r & -N_{\dot{r}}r_r & 0 & K_{\dot{p}}p_r \\ Y_{\dot{v}}v_r & -X_{\dot{u}}u_r & 0 & M_{\dot{q}}q_r & -K_{\dot{p}}p_r & 0 \end{bmatrix}. \quad (1.6)$$

$$\boldsymbol{D}(\boldsymbol{v}) = -[X_{u|u|}, Y_{v|v|}, Z_{w|w|}, K_{p|p|}, M_{q|q|}, N_{r|r|}]^{DT}. \quad (1.7)$$

2.4. Hydrodynamic parameters

Hydrodynamic parameters consist of the added mass and damping terms, as shown in equation (1.5), (1.6) and (1.7). The equation of added mass was derived based on the potential flow theory which is written as [11]

$$A_{ii} = \rho C_a V_R. \quad (1.8)$$

As potential flow theory does not consider the viscous effects, an empirical formula was introduced to include the effect of viscosity of the fluid which is called the damping or drag term. This can be written as [11]

$$B_{ii} = \frac{\rho}{2} C_D A_R. \quad (1.9)$$

These two parts of the equations together are called Morison's equation [12].

The hydrodynamic coefficient C_a and C_D in (1.8) and (1.9) for the bodies of different shapes are available in [11] which are tabulated based on DNV standards.

The hydrodynamic parameters of the Minerva HAUV were taken from [13] which were then modified for all the vehicles in the system to get their effect about O. The hydrodynamic parameters of the cubic payload and cylindrical links were calculated using the coefficients from the empirical data given in reference [11].

2.5. Hydrostatic parameters

The hydrostatic parameters depend on weight and buoyancy. In fact, the Minerva HAUV was designed such that its buoyancy is slightly higher than its weight to enable it to come to the surface in case of an emergency. However, for the purpose of this paper, the Minerva HAUVs were assumed to be neutrally buoyant. The solid links were also assumed to be neutrally buoyant. Conversely, the payload was selected such that the difference between its weight and buoyancy is within the thrust limit of the vertical thrusters on the combined system.

As aforementioned, all the force and moment terms were defined in BFF. Therefore, the hydrostatic parameters are required to be transformed into BFF. After transformation, the hydrostatic parameters become

$$\boldsymbol{g}(\boldsymbol{\eta}) = \begin{bmatrix} -(W-B)\sin\theta \\ (W-B)\cos\theta\sin\phi \\ (W-B)\cos\theta\cos\phi \\ z_b B\cos\theta\sin\phi \\ z_b B\sin\theta \\ 0 \end{bmatrix}. \quad (1.10)$$

z_b appears as COB is above the COG to ensure static stability. The hydrostatic terms which are required to calculate the hydrostatic parameters are given in Table 1.

Table 1: Hydrostatic terms

	Minerva HAUV	Link	Payload
Mass (Kg)	520	98	Needs to be worked out
Length (m)	1.5	2	1
Width (m)	0.82	-	1
Height (m)	0.84	-	1
Diameter (m)	-	0.25	-

3. PROPULSION MODEL

There are 5 thrusters on each Minerva HAUV i.e. two axial, two vertical and one transverse. The effect of each thruster is required to be taken about O.

The combined thrust vector can be written as a product of thrust allocation matrix $(\boldsymbol{T_a})$ and thrust force vector $(\boldsymbol{f})$, given as

$$\boldsymbol{\tau} = \boldsymbol{T_a f}. \quad (1.11)$$

A total of 20 thrusters are installed on the four Minerva HAUVs, as shown in Figure 3. Therefore, the thrust force vector becomes

$$\boldsymbol{f} = [f_1 \ f_2 \ f_3 \dots f_{20}]^{\boldsymbol{T}}. \quad (1.12)$$

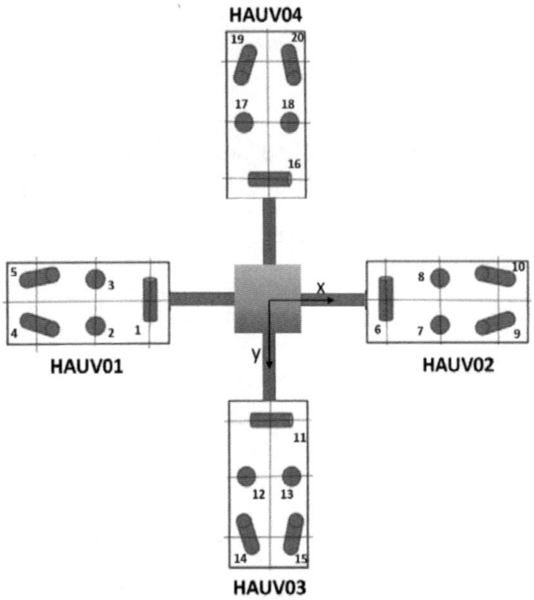

Figure 3: Allocation of thrusters in the combined system

The effect of each thruster on each vehicle is shown in the columns of the thrust allocation matrix. The thrust allocation matrices of the individual vehicles are given as

$$\boldsymbol{T_1} = \begin{bmatrix} 0 & 0 & 0 & c\sigma & c\sigma \\ 1 & 0 & 0 & s\sigma & -s\sigma \\ 0 & 1 & 1 & 0 & 0 \\ 0 & l_{y2} & l_{y3} & 0 & 0 \\ 0 & -l_{x2} & -l_{x3} & 0 & 0 \\ l_{x1} & 0 & 0 & l_{x4}c\sigma - l_{y4}s\sigma & l_{x5}c\sigma - l_{y5}s\sigma \end{bmatrix}. \quad (1.13)$$

$$\boldsymbol{T_2} = \begin{bmatrix} 0 & 0 & 0 & c\sigma & c\sigma \\ 1 & 0 & 0 & s\sigma & -s\sigma \\ 0 & 1 & 1 & 0 & 0 \\ 0 & l_{y7} & l_{y8} & 0 & 0 \\ 0 & -l_{x7} & -l_{x8} & 0 & 0 \\ l_{x6} & 0 & 0 & l_{x9}c\sigma - l_{y9}s\sigma & l_{x10}c\sigma - l_{y10}s\sigma \end{bmatrix}. \quad (1.14)$$

$$\boldsymbol{T_3} = \begin{bmatrix} 1 & 0 & 0 & -s\sigma & s\sigma \\ 0 & 0 & 0 & -c\sigma & -c\sigma \\ 0 & 1 & 1 & 0 & 0 \\ 0 & l_{y12} & l_{y13} & 0 & 0 \\ 0 & -l_{x12} & -l_{x13} & 0 & 0 \\ -l_{y11} & 0 & 0 & l_{x14}s\sigma - l_{y14}c\sigma & l_{x15}s\sigma - l_{y15}c\sigma \end{bmatrix}. \quad (1.15)$$

$$\boldsymbol{T_4} = \begin{bmatrix} 1 & 0 & 0 & -s\sigma & s\sigma \\ 0 & 0 & 0 & -c\sigma & -c\sigma \\ 0 & 1 & 1 & 0 & 0 \\ 0 & l_{y17} & l_{y18} & 0 & 0 \\ 0 & -l_{x17} & -l_{x18} & 0 & 0 \\ -l_{y16} & 0 & 0 & l_{x19}s\sigma - l_{y19}c\sigma & l_{x20}s\sigma - l_{y20}c\sigma \end{bmatrix}. \quad (1.16)$$

Where the position of each thruster from O is tabulated, as shown in Table 2.

Table 2: Position of thrusters in the combined system

Thruster	l_x**(m)**	l_y**(m)**	l_z**(m)**
Thruster 01	-2.55	0	0
Thruster 02	-2.75	0.2	0
Thruster 03	-2.75	-0.2	0
Thruster 04	-3.26	0.3	0
Thruster 05	-3.26	-0.3	0
Thruster 06	2.55	0	0
Thruster 07	2.75	0.2	0
Thruster 08	2.75	-0.2	0
Thruster 09	3.26	0.3	0
Thruster 10	3.26	-0.3	0
Thruster 11	0	2.55	0
Thruster 12	-0.2	2.75	0
Thruster 13	0.2	2.75	0
Thruster 14	-0.3	3.26	0
Thruster 15	0.3	3.26	0
Thruster 16	0	-2.55	0
Thruster 17	-0.2	-2.75	0
Thruster 18	0.2	-2.75	0
Thruster 19	-0.3	-3.26	0
Thruster 20	0.3	-3.26	0

The combined thrust allocation matrix ($\boldsymbol{T_a}$) becomes

$$\boldsymbol{T_a} = [\boldsymbol{T_1}\ \boldsymbol{T_2}\ \boldsymbol{T_3}\ \boldsymbol{T_4}]. \quad (1.\ 17)$$

HAUV 03 and 04 are placed at angles of 90° from HAUV 01 and 02 respectively. The axial thruster makes an angle of 10° with the horizontal (σ), therefore, it contributes towards both the surge and sway motions. However, due to the position of the vehicles, the thrust force produced by the axial thrusters in the combined system contribute equally in surge and sway. Moreover, the transverse thrusters on HAUV 01 and 02 apply in sway, whereas, on HAUV 03 and 04 thus apply in the surge direction. The detailed thrust allocation matrices are given in the Appendix.

4. SYSTEM RESPONSE WITHOUT CONTROL AND SEA CURRENT

In order to verify the simulation code and optimise the system for the payload mass, the motion response was first analysed without implementing the control system and ignoring the sea current effect.

The maximum mass of the payload which can be transported was first worked out analytically. The volume of the container which accommodates the payload is 1m^3. Therefore, the buoyancy force of the payload becomes 10055N. Each thruster on the Minerva HAUV can apply a maximum thrust force of 340N [14]. The eight vertical thrusters of the combined system produce a total vertical thrust of 2720N. Therefore, the net upward force becomes 12775N. If the weight of the payload is equal to 12775N, the system would hover at a vertical position. Hence, the mass of the payload must be lower than 1302Kg to ensure motion of the system in all directions.

Test 01: To compare with the analytical outcome, the payload mass was kept equal to 1302Kg in the simulation model. The motion response is shown in Figure 4. Though keeping the exact zero position is not possible in the simulation model due to the nonlinear coupled dynamics, it is verified that the mass of the payload should be lower than 1302Kg for controlled manoeuvring of the system in all directions.

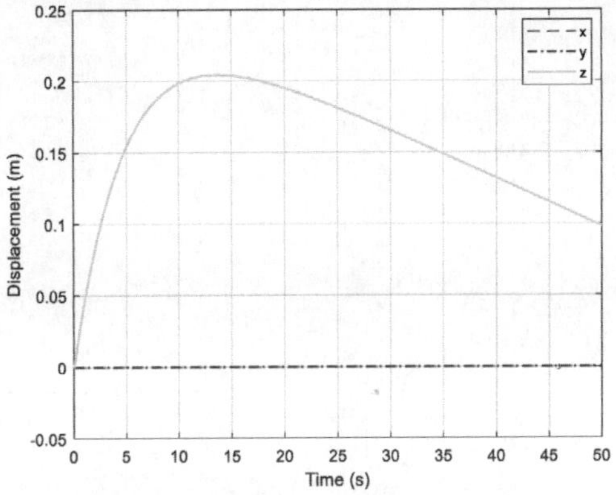

Figure 4: Test 01 - motion response

Test 02: The HAUVs were positioned in the combined system in such a way that their axial thrusters apply their positive thrusts inward i.e. towards the payload. Therefore, if all the axial thrusters are applied positively, there should not be any motion in the axial direction. In this test, the mass of the payload was kept at its maximum and the vertical thrusters were applied at their maximum thrust to keep the vertical position. The axial thrusters were applied at their maximum positive thrust. From Figure 5, a small vertical motion of around 0.2m is observed before it comes down to zero, similar to the motion response in Figure 4. However, the motion in the surge is zero.

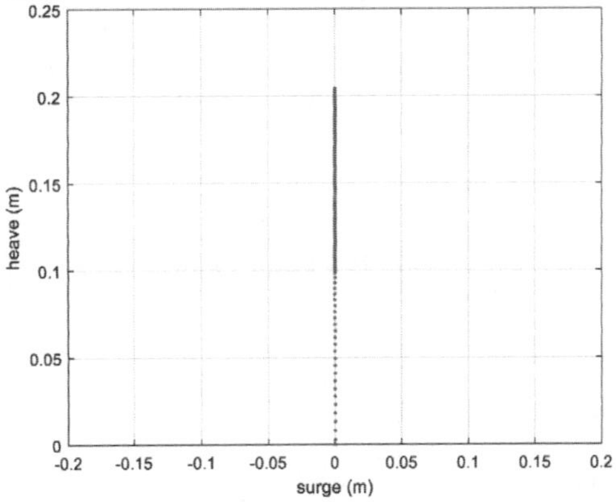

Figure 5: Test 02 - motion response

Test 03: This test was designed to observe the motion response of the system when all the transverse thrusters are applied. Therefore, a 50N thrust force was applied on the four transverse thrusters of the four HAUVs. The vertical thrusters were also applied to maintain the vertical position. The transverse thrusters of HAUV 01 and 02 act in the y-direction, whereas, of HAUV 03 and 04 in the x-direction. Therefore, a thrust force of 100N is experienced in each x and y directions, as shown in Figure 6. The damping force increases until the system is accelerating and comes to a constant value once the uniform velocity is reached. The system moves an equal distance in surge and sway, as shown in Figure 7.

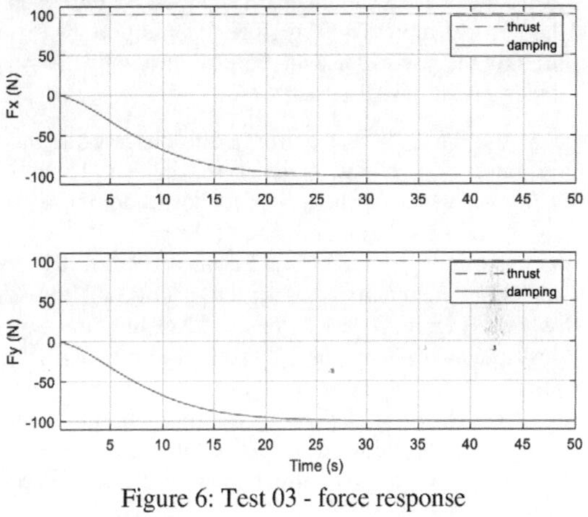

Figure 6: Test 03 - force response

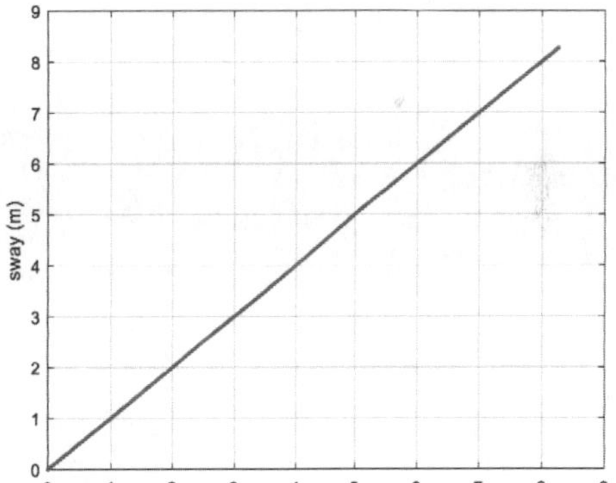

Figure 7: Test 03 - motion response

5. SYSTEM RESPONSE WITH SEA CURRENT

The sea current when applied opposite to the motion of the system increases the resistive force, however, when applied in the direction of the motion, assists the motion. It can even divert the system from its path if applied at an angle to the direction of motion.

The sea current velocity depends on depth and area of operation. It can rise to 3000cm/sec in the strait of Gibraltar. Generally, it is measured in the range of 1 to 20 cm/sec [15].

It was initially necessary to work out if in each direction the system would be able to overcome the sea current disturbances. The axial thrusters due to their installation at an angle produce some amount of their thrust in the transverse direction. Due to the position of HAUV 03 and 04 at an angle 90 degrees with HAUV 01 and 02, the axial thrusters apply equal and half of their thrust force (1360N) in each surge and sway. Furthermore, two of the transverse thrusters act in the y-direction and two in the x-direction. This makes a total of 2040N thrust force in each surge and sway. In the vertical direction, there is a total of 8 thrusters which can produce a maximum thrust force of 2720N. However, the vertical thrusters have to balance out the difference between weight and buoyancy of the system, to overcome the sea current effect and to move the system in the vertical direction. Therefore, it is important to set a limit on the payload mass to be easily lifted by the system under sea current disturbances.

Test 04: This test was performed to observe the motion response of the system under sea current velocity of 0.2m/sec when applied right opposite to the vertical motion of the system. The mass of the payload was kept at its maximum i.e. 1302Kg and each vertical thruster was applied at the maximum upward thrust of 340N. It can be seen in Figure 8 that instead of keeping the vertical position, the system moves down in the vertical direction. This is because the sea current increases the resistive force by 200N when applied at a velocity of 0.2m/sec in the direction opposite to the direction of motion. Hence the resistive force of sea current reduces the load carrying capacity of the system.

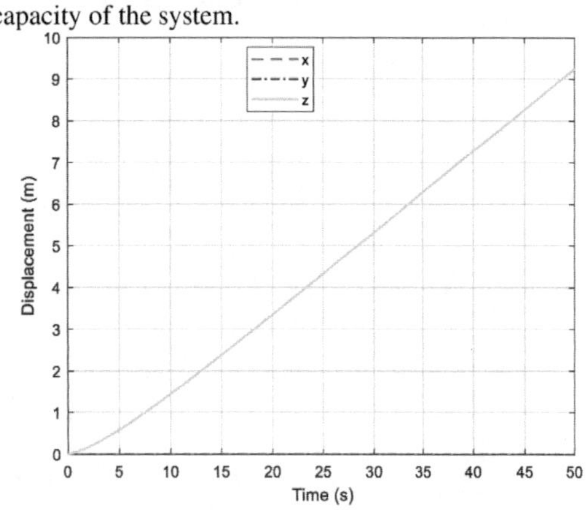

Figure 8: Test 04 – displacement response

Test 05: From test 04, it was found that a sea current velocity of 0.2m/sec applies a resistive force of 200N. Therefore, the mass of the payload was reduced by 20.4Kg to balance the forces. This makes the total mass of the payload equal to 1281.6Kg. Ideally, there should not be any motion. However, the resistive force of the sea current induces a constant velocity of around 0.06m/sec, which moves the system vertically down, as shown in Figure 9 and Figure 10 respectively. Therefore, the mass of the payload is required to be further reduced to maintain the vertical position. This helps in deciding the maximum payload which could be feasibly transported.

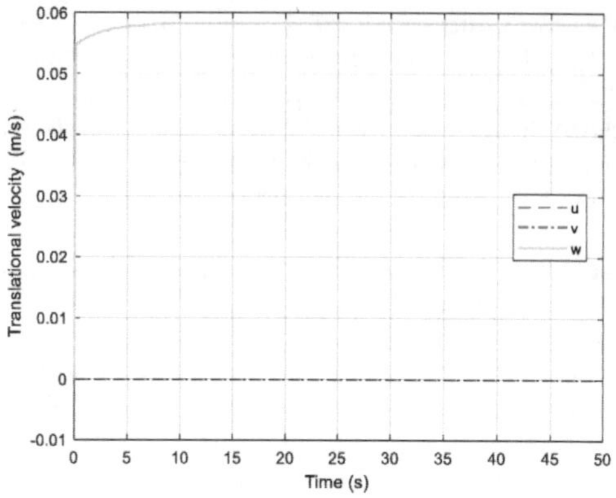

Figure 9: Test 05 – translational velocity response

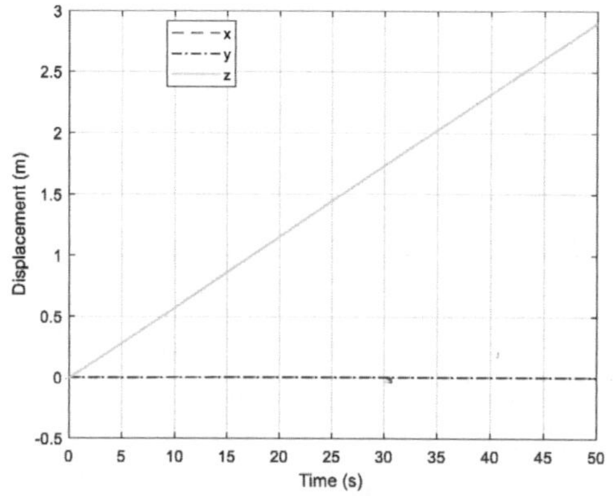

Figure 10: Test 05 – displacement response

Test 06: In this test, the payload mass was reduced to 1269Kg. This results in an increase of upward velocity produced by the vertical thrusters which are operated at their maximum thrust to overcome the induced velocity by the sea current, as shown in Figure 11. The system now maintains the vertical position, as shown in Figure 12.

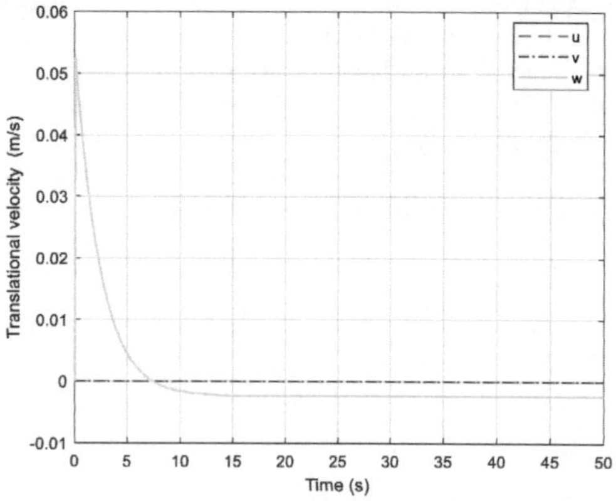

Figure 11: Test 06 – translational velocity response

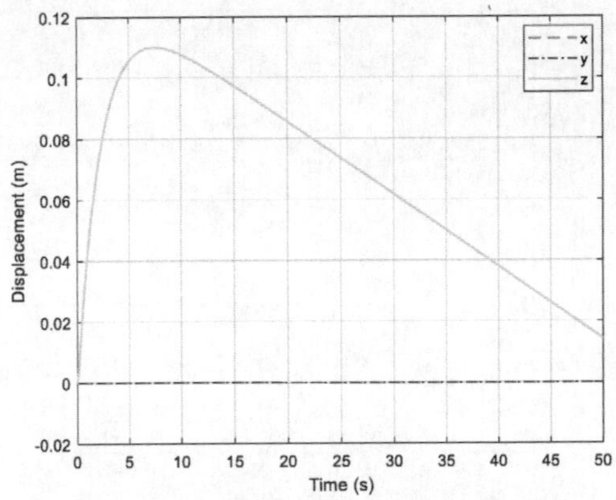

Figure 12: Test 06 – displacement response

It is concluded that the system can hover at the maximum payload mass of 1269Kg under the opposing sea current velocity of 0.2m/sec. However, to move the system up vertically, the mass of the payload should be lower than 1269Kg. Therefore, 1200Kg was set as a feasibly safe payload mass under such sea current conditions.

Test 07: For this test, the mass of the payload was reduced to 1200Kg. The sea current velocity of 0.2m/sec was now applied in the axial direction while all the vertical thrusters were applied to move the system up in the vertical direction. This results in both the surge and heave displacements as well as the pitch rotation, as shown in Figure 13 and Figure 14 respectively.

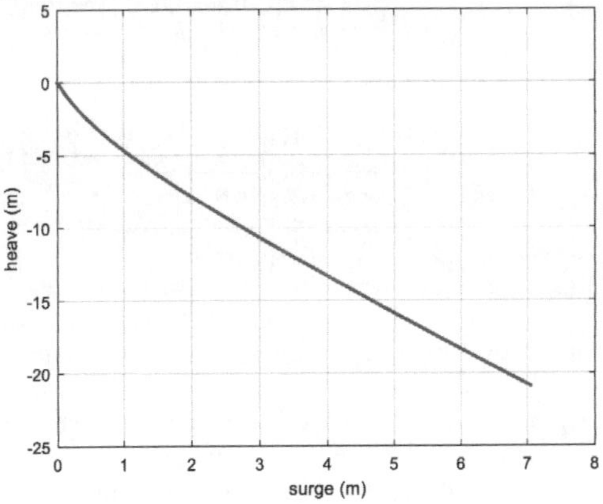

Figure 13: Test 07 – displacement response

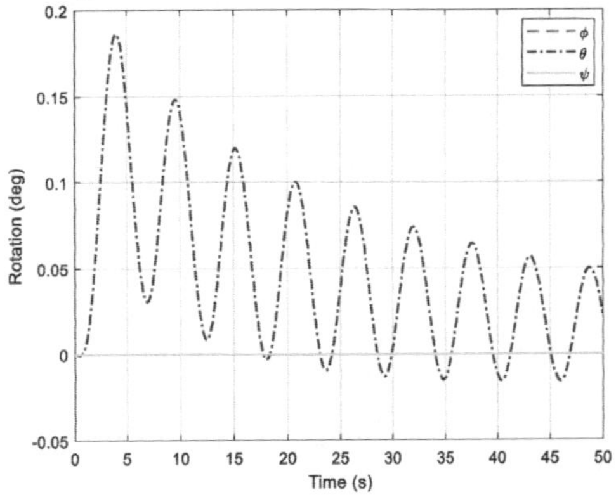

Figure 14: Test 07 – rotation response

6. CONTROL SYSTEM DESIGN

In order to control the system response under sea current effects, a control system was designed with the PID controllers, as shown in Figure 15.

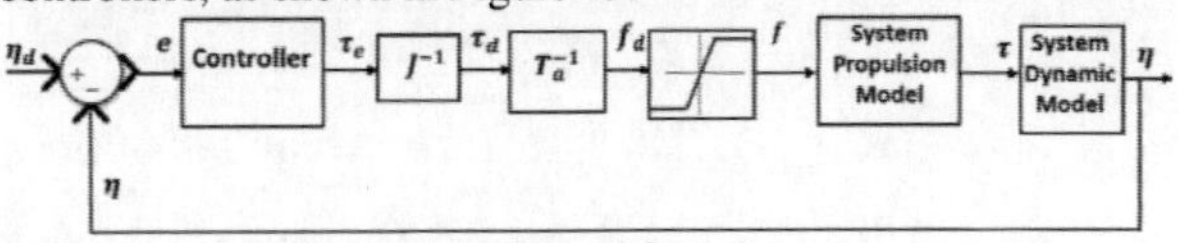

Figure 15: Control System

PID controllers are applied in all the directions even if no motion is desired in a direction. This is because when the system moves to the desired position in one direction it could produce motions in the other directions due to the coupling effect. Therefore, it is necessary to apply controllers in the other directions to keep their values to zero.

The payload mass is kept at 1200Kg to ensure manoeuvring in all directions as analysed earlier.

Test 08: The PID controllers were applied to move the system to a desired distance of 10m in surge against the sea current velocity of 0.2m/sec. It is assumed that the vertical thrusters have balanced out the hydrostatics to maintain a vertical position.

The PID gains are tabulated in Table 3.

Table 3: Test 08 – PID gains

	K_p	K_i	K_d
Surge	200	0.1	0.1
Other directions	3	0.1	0.1

From Figure 16, the resistive damping force increases to 1650N due to the effect of sea current. Therefore, a thrust force of 2000N was applied by the axial thrusters to overcome it and to move the system in surge, as shown in Figure 17. It is observed that the applied thrust and the opposing damping forces gradually reduce in value as the system approaches the desired position. This is very important as if the forces suddenly drop to zero when the system reaches the desired position would result in undesirable oscillations. Finally, when the desired position is reached, an axial thrust force is constantly applied to overcome the continually applied sea current.

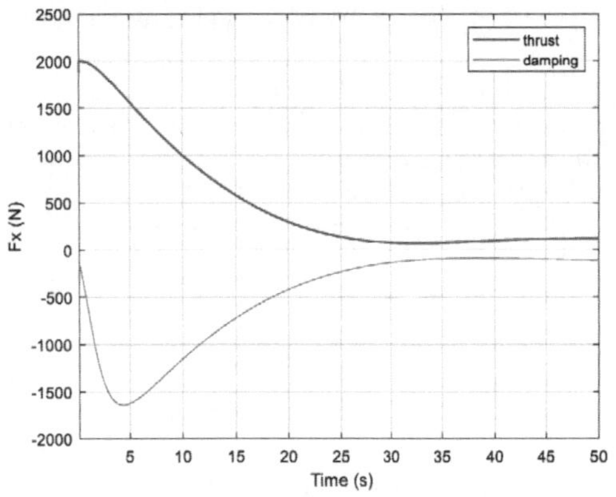

Figure 16: Test 08 – force response

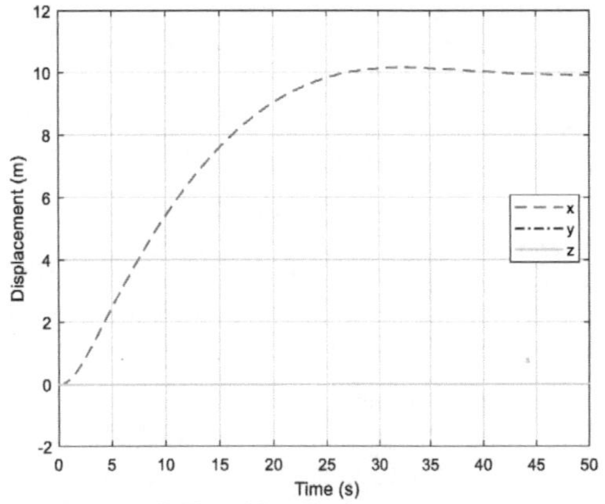

Figure 17: Test 08 – displacement response

Test 09: The critical situation arises when the sea current applies in the direction other than the direction of motion of the system. In this test, for the desired axial distance of 10m, the sea current velocity of 0.2m/sec was applied in the vertical direction. Keeping the same PID gains as in test 08, the system experiences the pitch oscillations, as shown in Figure 18 and an undesirable motion in heave, as shown in Figure 19.

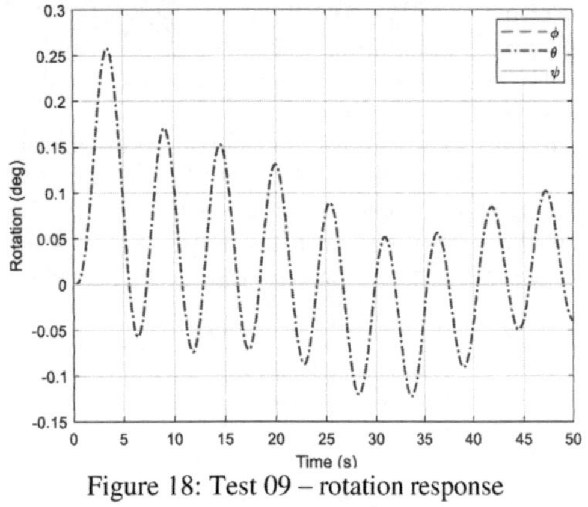

Figure 18: Test 09 – rotation response

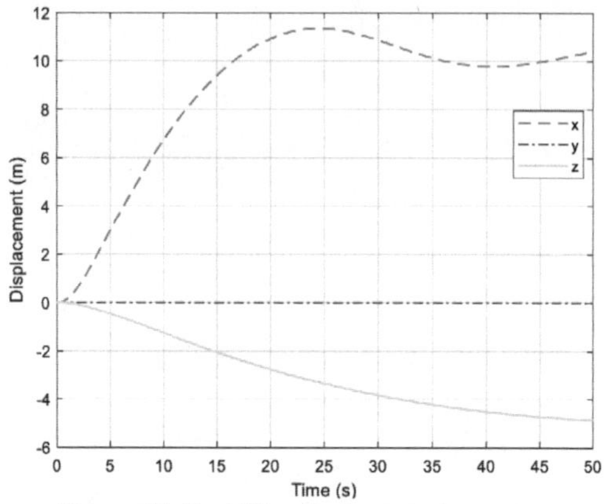

Figure 19: Test 09 – displacement response

Test 10: For the same condition as in test 09, the PID gains were modified to overcome the disturbances caused by the sea current applied perpendicular to the direction of motion, as shown in Table 4. The gains are increased in surge, heave and pitch to cover the desired surge distance and overcome the disturbances caused by the vertical sea current.

Table 4:Test 10 – PID gains

	K_p	K_i	K_d
Surge	300	0.1	200
Sway	3	0.1	0.1
Heave	200	2	2
Roll	3	0.1	0.1
Pitch	100	1	100
Yaw	3	0.1	0.1

From Figure 20, the desired surge position is reached by the system with small offset which is acceptable. The heave position is also brought to zero after the initial oscillation. From Figure 21, the pitch rotation is approaching zero degrees over time but with abrupt oscillations. This is not acceptable as to overcome these oscillations, constant opposite pitch moment is applied by the thrusters which could result in the thrusters to get malfunctioned. Moreover, the PID gains are very high especially the differential gain. This makes the system sensitive to noise.

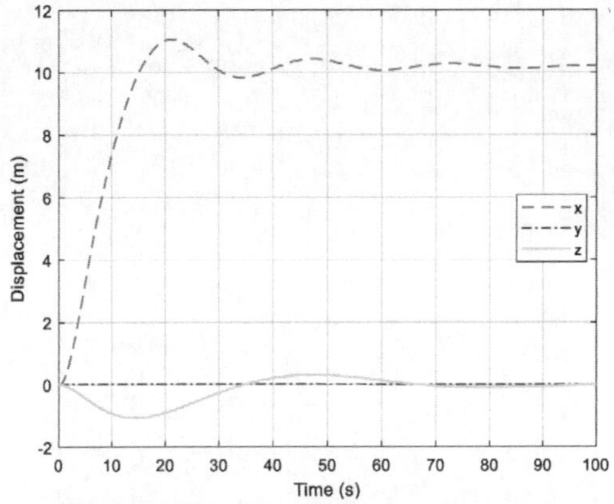

Figure 20: Test 10 – displacement response

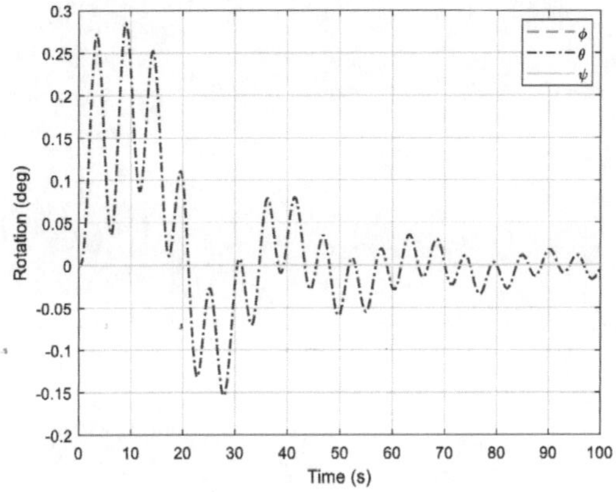

Figure 21: Test 10 – rotation response

7. CONCLUSIONS

The analysis in this paper focuses on the impact of sea current on the motion response of a system of four Minerva HAUVs which are rigidly attached to a cubic payload via solid cylindrical links. The simulation model was first analysed without implementing the control system to set a limit on the payload mass which can be safely transported. The maximum payload mass at which the system would hover without the sea current effect was worked out to be 1302Kg, which was reduced to 1269Kg when the sea current was applied at a velocity of 0.2m/sec right opposite to the direction of motion of the system. Thus, the payload mass of 1200Kg was decided to be a feasible option for manoeuvring under the sea current velocity of 0.2m/sec. The limit on payload mass would further reduce with the increase of sea current velocity. A control system was designed with the PID controllers to overcome the disturbances caused by the sea current and to reach the desired location. The control system worked satisfactorily when the sea current was applied right opposite to the direction of the motion keeping in consideration the payload and sea current velocity limits. On the other hand, the control system had to work harder with higher PID gains to get the desired motion response when the sea current was applied normal to the direction of motion. However, it is not acceptable as abrupt pitch oscillations were not overcome which could result in the malfunctioning of the thrusters. Moreover, the higher PID gains especially the differential gain makes the system sensitive to the noise. It is concluded that the PID controller has limitations to control the motion response when a sea current is applied on the system at an angle to the direction of motion. It is suggested that the system performance is analysed with the optimal controllers and investigates the system performance at a higher level of sea current in the future.

8. REFERENCES

[1] J. Kennedy, A. A. Proctor, E. Gamroth, and C. Bradley, "Development of a Highly Autonomous Underwater Vehicle for Scientific Data Collection," pp. 1–11.

[2] D. R. Blidberg, "The Development of Autonomous Underwater Vehicles (AUV); A Brief Summary," *IEEE ICRA*, vol. 6500, p. 12, 2010.

[3] A. Phillips, M. Furlong, and S. R. Turnock, "The use of Computational Fluid Dynamics to Determine the Dynamic Stability of an Autonomous Underwater Vehicle," *10th Numer. Towing Tank Symp. (NuTTS'07), Hamburg, Ger.*, 2007.

[4] C. Chen *et al.*, "Computational fluid dynamics study of the motion stability of an autonomous underwater helicopter," *Ocean Eng.*, pp. 1–13, 2017.

[5] J. Ghommam, H. Mehrjerdi, M. Saad, and F. Mnif, "Formation path following control of unicycle-type mobile robots," *Rob. Auton. Syst.*, vol. 58, no. 5, pp. 727–736, 2010.

[6] D. Mellinger, "Trajectory Generation and Control for Quadrotors," Mechanical Engineering and Applied Mechanics PhD Thesis, University of Pennsylvania, 2012.

[7] N. Michael, J. Fink, and V. Kumar, "Cooperative manipulation and transportation with aerial robots," *Auton. Robots*, pp. 1–14, 2010.

[8] T. I. Fossen, *Handbook of Marine Craft Hydrodynamics and Motion Control*, 1st Ed. Sussex: John Wiley & sons, 2011.

[9] F. U. Rehman, G. Thomas, and E. Anderlini, "Centralized Control System Design for Underwater Transportation using two Hovering Autonomous Underwater Vehicles (HAUVs)," *IFAC-PapersOnLine*, vol. 52, no. 11, pp. 13–18, 2019.

[10] J. Peraire and J. Widnall, "Lecture L26 - 3D Rigid Body Dynamics: The Inertia Tensor," in *Dynamics*, 2008.

[11] DNV, "Modelling and Analysis of Marine Operations," *tech. rep. DNV-RP-H103*, pp. 1–91, 2010.

[12] J. R. Morison, M. P. O'Brien, J. W. Johnson, and S. A. Schaf, "The force exerted by surface waves on piles," *Pet. Technol.*, vol. 2, no. 5, 1950.

[13] S. Marie Mo, "Development of a Simulation Platform for ROV systems," Marine Technology Master Thesis, Norwegian University of Science and Technology, 2015.

[14] C. Christo, "Design of Dynamic Positioning System for ROV Minerva," 2010.

[15] G. Shanmugam, "New Perspectives on Deep-water Sandstones," in *Handbook of Petroleum Exploration and Production*, 2012.

9. AUTHORS BIOGRAPHY

Faheem Ur Rehman is currently undertaking PhD studies at UCL Mechanical Engineering department. His field of research is autonomous underwater transportation using multiple autonomous vehicles. He is a Member of the Royal Institute of Naval Architects (RINA) and a student member of the Society of Naval Architects and Marine Engineers (SNAME).

Dr Enrico Anderlini is a senior research associate at UCL Mechanical Engineering. He teaches ocean engineering at undergraduate and postgraduate level and provides support to the ship and submarine design exercises. During his doctorate, Enrico developed control strategies for wave energy converters using reinforcement learning. He is an Associate Member of the Royal Institute of Naval Architects (RINA).

Professor Giles Thomas is BMT Chair of Maritime Engineering in University College London's Department of Mechanical Engineering. A naval architect, his research focusses on the performance of ships, boats and offshore structures. Specific interests include fluid-structure interaction, hydrodynamics, full-scale measurements, model testing and design. He is a Fellow of the Royal Institution of Naval Architects (RINA) and a Chartered Engineer (CEng – Engineering Council of Great Britain).

10. APPENDIX

$$\boldsymbol{T_1} = \begin{bmatrix} 0 & 0 & 0 & 0.98 & 0.98 \\ 1 & 0 & 0 & 0.17 & -0.17 \\ 0 & 1 & 1 & 0 & 0 \\ 0 & 0.2 & -0.2 & 0 & 0 \\ 0 & 2.75 & 2.75 & 0 & 0 \\ -2.55 & 0 & 0 & -0.86 & 0.86 \end{bmatrix}. \quad \text{(A.1)}$$

$$\boldsymbol{T_2} = \begin{bmatrix} 0 & 0 & 0 & -0.98 & -0.98 \\ 1 & 0 & 0 & 0.17 & -0.17 \\ 0 & 1 & 1 & 0 & 0 \\ 0 & 0.2 & -0.2 & 0 & 0 \\ 0 & -2.75 & -2.75 & 0 & 0 \\ 2.55 & 0 & 0 & 0.86 & -0.86 \end{bmatrix}. \quad \text{(A.2)}$$

$$\boldsymbol{T_3} = \begin{bmatrix} 1 & 0 & 0 & -0.17 & 0.17 \\ 0 & 0 & 0 & -0.98 & -0.98 \\ 0 & 1 & 1 & 0 & 0 \\ 0 & 2.75 & 2.75 & 0 & 0 \\ 0 & 0.2 & -0.2 & 0 & 0 \\ -2.55 & 0 & 0 & 0.86 & -0.86 \end{bmatrix}. \quad \text{(A.3)}$$

$$\boldsymbol{T_4} = \begin{bmatrix} 1 & 0 & 0 & -0.17 & 0.17 \\ 0 & 0 & 0 & 0.98 & 0.98 \\ 0 & 1 & 1 & 0 & 0 \\ 0 & -2.75 & -2.75 & 0 & 0 \\ 0 & 0.2 & -0.2 & 0 & 0 \\ 2.55 & 0 & 0 & -0.86 & 0.86 \end{bmatrix}. \quad \text{(A.4)}$$

PAVING THE WAY TOWARD AUTONOMOUS SHIPPING DEVELOPMENT FOR EUROPEAN WATERS – THE AUTOSHIP PROJECT

V. Bolbot, G. Theotokatos, E. Boulougouris, University of Strathclyde, UK
L.A. L. Wennersberg, H. Nordahl, Ø. J. Rødseth SINTEF, Norway
J. Faivre, BUREAU VERITAS Marine & Offshore, France
M. Molica Colella CiaoTech (PNO Group), Italy

SUMMARY
New developments in maritime industry include the design and operation of autonomous ships. The AUTOSHIP project is one initiative promoting the use of autonomous ships in European waters focusing on two specific use cases, a Short Sea Shipping (SSS) cargo vessel and an Inland Waterways (IWW) barge. The AUTOSHIP objectives include thorough regulatory, societal, financial, safety and security analyses for the two investigated use cases as well as the development of a novel framework and methods for the design of autonomous vessels. This objective is achieved with the support of a number of activities, including supply chain, regulatory, risk and gaps analyses. Some results and findings from these activities are presented in this paper. The results demonstrate that the supply chain analysis is important to understand the complex relationships between different partners and phases for the effective design of maritime autonomous systems. Furthermore, a number of regulatory gaps needs to be addressed for the wider adoption of the AUTOSHIP use cases. There is a number of essential hazards associated with each of the two use cases; measures to mitigate these hazards are presented.

1. INTRODUCTION

Paving the way towards the development of the Maritime Autonomous Surface Ships (MASS) requires innovative and bold initiatives. Such a collaborative initiative is the AUTOSHIP project [1], which aims at converting two conventional vessels in remotely controlled and autonomous vessels as well as testing and operating them; in particular, a Short Sea Shipping (SSS) cargo vessel and an Inland Waterways (IWW) barge. AUTOSHIP is a €27M budget project funded by EU with a duration of 42 months, which started on 1st June 2019. The AUTOSHIP consortium involving 10 partners gathers the technological excellence as well as the market, industry and academic knowledge and know-how to analyse and define the extended framework related to MASS adoption and acceptance.

The main aim of AUTOSHIP project is the promotion of autonomous shipping in Europe. The project specific objectives focus on the development of key enabling technologies, including shore/remote control centres, autonomous navigation, intelligent maintenance, operation and safety based on Artificial Intelligence algorithms. In addition, communication technology enabling a prominent level of cyber security, will be developed and experimentally validated targeting both retrofitting of existing vessels and new ships design.

Moreover, another project objective is the development and use of new tools including simulators and decision support tools for operators training, cost assessment and optimisation of the next generation autonomous ships. It is expected that the developed tools and methods will support the operation and optimisation of conventional vessels as well. The project objectives also include the review of the environmental and socio-economic aspects for the autonomous shipping transportation and scaling up at a global level. This will allow for quantification of societal, environmental and financial impact stemming from modal-shifts from road to waterborne transport, jobs shifting towards higher-skills positions, and new business cases.

The safety and security (including cyber-security) is another objective of paramount importance for the development of MASS. Two work packages led by the Maritime Safety Research Centre (MSRC) of the University of Strathclyde [2] are dedicated to activities, which include comprehensive regulatory, societal, economic, (cyber)security and safety analyses, thus leading to the development of a roadmap, methodologies, advanced methods and standards for planning and design of next generation MASS, duly motivated by the two full-scale demonstrators testing. Based on the project results a proposal to IMO for amending and improving the existing regulatory framework is planned.

The realisation of the last-mentioned objective of the AUTOSHIP project is implemented according to the following steps. First, the two investigated vessels supply chains are analysed including the operating phases and the involved actors. Subsequently, the existing regulatory framework for the two use cases is analysed. Based on the available information for the two use cases, a risk assessment is implemented and risk control measures are defined. These investigations support the development of a novel safety, security and cybersecurity framework focusing on autonomous ships. Based on these steps the major gaps for scaling-up the demonstrators are provided. These steps are elaborated further in the next sections and some of the project findings are presented.

2. SUPPLY CHAIN ANALYSIS

This is a task led by SINTEF [3], aiming at the supply chain analysis to provide the foundation for following-up analyses on: (a) the impact of autonomy introduction on regulations; (b) the societal and economic scenarios; (c) definition of the basic operational scenarios for the risk assessment. This analysis also provides input for determining the operations, functions and controls for the two use cases along with the vessels routes, including segments with remote controlled/monitored sailing and totally unmanned bridge.

This study and analysis are based on information from semi-structured interviews carried out with staff from the AUTOSHIP project operators, specifically, Blue Line Logistics [4] and Eidsvaag [5]. The survey was built up to capture all stages of the investigated vessels supply chains according to the multi-stage supply chain model described in [6]. Emphasis was put on distribution and logistics chain, but also to capture relevant interactions with production planning and inventory control. The results of the semi-structured interviews were then used to map the supply chains by using flowchart models that describe both the sequence and relationships between the flow of goods and the flow of information.

Some indicative results for the IWW use case are provided in the flowchart of Figure 1. As it is deduced, the supply chain includes several actors: customers, sales department, logistics department, the transporting truck and IWW manager. The transportation by ship is arranged only after a coordination between the land transport and ship transport. This understanding is important for the financial analysis. The results of the detailed supply chain analysis are available in the public project deliverable [1].

3. COMPLIANCE FRAMEWORK ANALYSIS

The AUTOSHIP project use cases are considered to operate in a specific regulatory and legislative environment, one in the inland waterways in Belgium and one in national waters as well as in a route between Norway and Denmark. The introduction of autonomous ship operations requires the understanding of the gaps that exist in the regulatory framework.

This need is addressed through task led by Bureau Veritas (BV) [7]. This task identifies and maps the existing regulatory framework including the prevailing regulations, rules and standards (international and national) for the SSS and IWW use cases considered in the AUTOSHIP project.

A systematic review approach and direct interaction with the relevant partners have been employed for identification and analysis of all the relevant information/material. This has allowed for mapping the compliance framework of the two use cases design and operation against the existing rules, regulations, standards and codes.

The analysis has been limited to mandatory instruments related to maritime safety and maritime security:

- Safety Of Life At Sea (SOLAS), International Convention on Load Lines (CCL), Tonnage, International Convention on Standards of Training, Certification and Watchkeeping (STCW), Convention on the International Regulations for Preventing Collisions at Sea (COLREG), Safety And Rescue (SAR) operations, Maritime Labour Convention (MLC), European Directives, National and Local regulations.
- Policy Regulations for the Navigation of Rhine (RPNR), European Code for Inland Waterways (CEVNI), Strasbourg Convention on the limitation of liability in inland navigation (CLNI), the Strasbourg Convention on the collection, deposit and reception of waste (CDNI), European Directives, Regional, National and Local regulations for the IWW use case.

IMO Interim Guidelines for MASS trials, IMO Guidelines on maritime cyber risk management and Bureau Veritas Guidelines for Autonomous Shipping [8] have also been considered. It should be noted that it does not include all codes and standards relevant for such SSS & IWW use cases (e.g. land based regulations for remote control centre located at shore).

The main identified areas where amendments or new developments are required are the following:

- Definition and responsibilities of the master, crew and responsible person in new context.
- Regulations requiring the presence and actions of human operators on-board, manual operations or indication/alarm on the bridge.
- No provisions for compulsory systems, devices and procedures that would facilitate crewless vessel operation (all tasks will be not moved from crew to

Figure 1 IWW supply chain overview

remote operators, some tasks will be performed by systems on board).
- Functions, rights and responsibilities of Remote Control Centre, including personnel qualification.

More detailed results of regulatory framework analysis are provided in the pertinent deliverable [1].

4. RISK ASSESSMENTS

The introduction of autonomous ships is accompanied with a number of challenges related to safety, security and cybersecurity. The safety challenges can be attributed to increased complexity related to the interactions between the autonomous ships systems/components and environment [9]. Furthermore, cybersecurity has been an important issue, as a cyber-attack can exploit vulnerabilities in the communication links and directly affect the integrity or availability of the data and control systems, leading to accidents [9, 10]. A number of incidences with unauthorised people gaining remote access to the ship control systems was recently reported [11]. Terrorists or pirates could potentially board an autonomous vessel trying to take control, which might lead to hazardous/inadvertent situations (e.g. collision with another ship or requests for significant ransom).

The availability of different methods for safety, security and cybersecurity analysis might create a confusion on their suitability and applicability as well as for the required sequence. For the purposes of this analysis, a classic HAZID as proposed according to the BV rules [8] with some modifications was selected. In addition, alternative approaches employed in other industries and their marinisation (customisation to the requirements of the maritime industry) are investigated. These include the SAE ARP4761 standard for development of civil aircrafts and systems as well as the ISO/PAS21488 and the approach of Safety of Indented Functionality (SOTIF). The risk assessment focuses mainly on the autonomous vessels and associated remote control centre functions during different operational phases. This task is primarily led by MSRC in close cooperation with BV and SINTEF.

The identification process focusing on system parts as shown in Figure 2 is implemented following the process presented in **Error! Reference source not found.**. First a specific group of functions is selected for the analysis. For this group of functions, a specific operating phase is considered. For each combination of functions and operating phases, a specific type of inadvertent event is considered either primarily related to safety, security or cybersecurity. The inadvertent event is identified by focusing on the output of a specific function. For identifying the potential hazards, guidance statements are used, such as provided wrong, not provided, provided in timely, provided results in conflict. For security/cybersecurity related events, different threat groups are considered. For each scenario, potential causes are considered in terms of safety (system error, failure, human error, inadvertent environment conditions, missing input), security (management failure, missing/faulty technical barrier, operating in dangerous area) and cybersecurity (potential cyberattacks types and vulnerabilities). For each cause and hazard, incident/accident categories are defined and potential consequences are identified in terms of safety, damage to environment or financial impact.

The main identified inadvertent events include the following:

- The surrounding situation is not properly (wrongly or not) determined e.g. Objects (small objects, navigation marks, ships, ship lights, floating objects, depth) are not detected and recognised, weather is not properly determined.

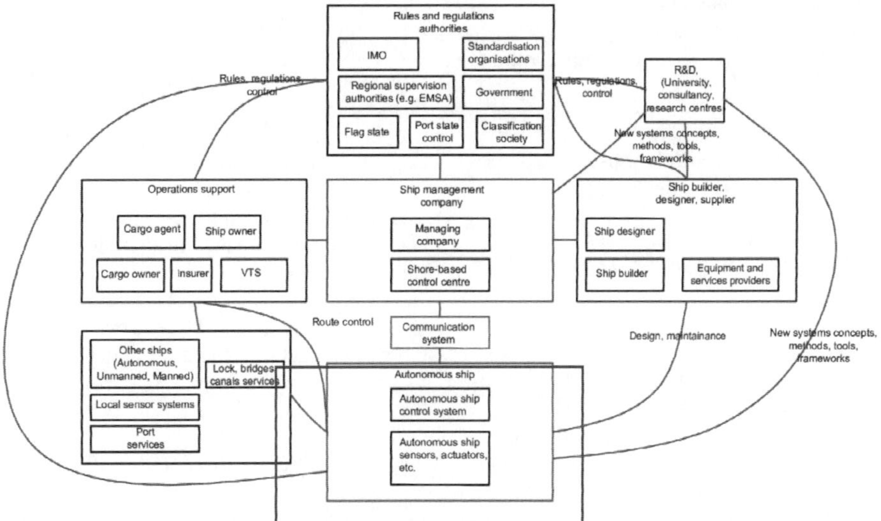

Figure 2 Analysis focus.

- Situation awareness equipment malfunction due to a generic cyberattack during berthing / sailing / manoeuvring.
- Ship on collision track with other vessels/on grounding track (no output / wrong output from collision avoidance system).
- Selecting route with heavy traffic/ bad weather conditions/ Close to shallow water/low visibility / close to obstacles.
- Vessel erratic operation/movements.

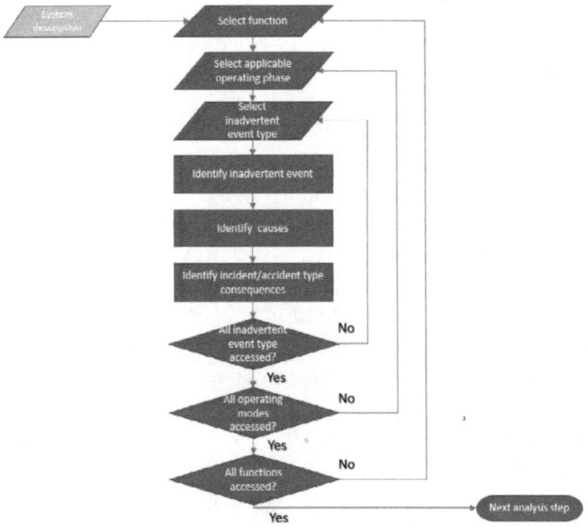

Figure 3 Risk analysis approach.

The potential risk mitigation measures include:

- Crosscheck of information coming from different sources using voting mechanisms.
- Installation of firewall and intrusion detection system in ship control network.
- Automatic remote-control centre personnel in case of serious malfunctions and take over.
- Double verification of selected route.
- Filters installed on sensors used for ship position control.

The results of the HAZID session along with detailed information for the risk and control measures will be provided in future publications.

5. NOVEL SAFETY, SECURITY, CYBERSECURITY ASSURANCE FRAMEWORK DEVELOPMENT

The MASS can be described as systems of Cyber-Physical Systems (CPSs). The additional complexity of the marine CPS on autonomous ships can be attributed to the novel components such as the shore control centre, the novel communication links, the novel human-machine interfaces, the extensive cyber (software) supported functions. It is expected that the existing safety assurance methods may have inherent weaknesses for supporting identification of all the potential hazardous scenarios [9, 12].

In this task, which is led by MSRC, a critical review of the methods and standards used for safety, security and cybersecurity assurance in different domains was implemented. Based on the standards and methods limitations, a novel safety assurance framework is proposed to support the design of safe, secure and cyber secure autonomous ships. The framework includes parallel but interacting processes for safety, security and cybersecurity of the MASS assurance. Each (safety, security, cybersecurity) assurance is double staged, proceeding from high-level analysis to lower, more detailed analysis and in parallel with Vee design process, as shown in
Figure 4. The analysis results are also used to derive specific test scenarios for MASS and support the development of the safety, security and cybersecurity case, which is addressed in other project tasks. In addition, a number of novel and advanced safety, security and cybersecurity analysis methods such as STPA [13], Cyber-Risk Assessment for Marine Systems [14] fitting into the new safety assurance framework is proposed for overcoming the existing methods drawbacks.

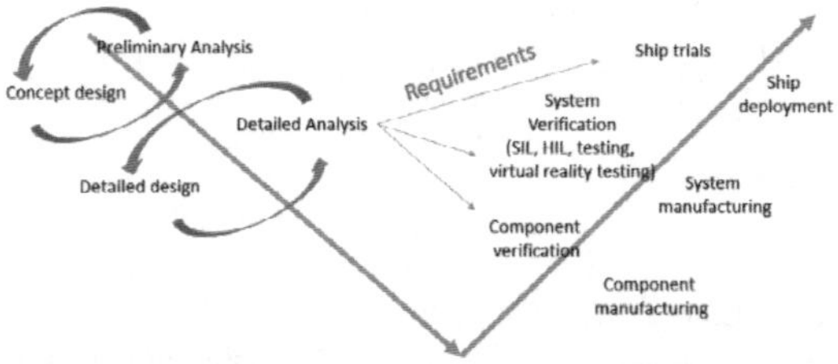

Figure 4 Proposed safety framework aligned with Vee design process.

6. GAP ANALYSIS

Using the output from previous tasks, the final purpose of this task led by MSRC is to identify gaps for wider adoption of the two demonstrators/use cases. This is carried out by implementing a multi-dimensional analysis.

The analysis is articulated in two steps. First the functional and operational requirements and Key Performance Indicators (KPIs) are defined considering the whole supply chain. Then, impact analysis of the identified scenarios is implemented for the two use cases. This is still work in process and results are anticipated to be published in the future studies.

7. CONCLUDING REMARKS

In this paper, the main objectives and some results of the AUTOSHIP project were presented. The results demonstrate that for the two use cases:

- The operation is implemented using a complex and highly interacting supply chain including several phases.
- A number of gaps exists in the existing regulatory framework, which need to be addressed to allow wider adoption of the use cases.
- Emerging hazards on the two use cases need to be addressed using appropriate control measures.
- The complexity of the autonomous ships requires the development of novel integrated design, safety, security, cybersecurity frameworks employing modern and advanced methods and tools for these systems analysis.
- The enhancement of operations and adoption for the novel use cases requires the definition of KPIs connected to the supply chain.

AUTOSHIP is an ongoing project and novel outputs are being constantly generated, such as methods, tools, concepts, results to support the design and operation of the two use cases. It is expected that these results will also support the general community of autonomous ships designers.

8. ACKNOWLEDGEMENTS

The study was carried out in the framework of the AUTOSHIP project (AUTOSHIP, 2019), which is funded by the European Union's Horizon 2020 research and innovation programme under agreement No. 815012. The authors greatly appreciate the AUTOSHIP partners reviewed and provided feedback for this paper. The authors affiliated with MSRC also greatly acknowledge the funding from DNV GL AS and RCCL for the MSRC establishment and operation. The opinions expressed herein are those of the authors and should not be construed to reflect the views of EU, DNV GL AS, RCCL or other involved partners in the AUTOSHIP project.

9. REFERENCES

1. AUTOSHIP. Autonomous Shipping Initiative for European Waters 2019 [Available from: https://www.autoship-project.eu/.
2. MSRC. The Maritime Safety Research Centre 2020 [Available from: https://www.strath.ac.uk/research/maritimesafetyresearchcentre/.
3. SINTEF. SINTEF 2020 [Available from: https://www.sintef.no/en/.
4. Blue Lines Logistics. Blue Lines Logistics News 2015 [Available from: http://www.bluelinelogistics.eu/news.
5. Eidsvaag. Eidsvaag 2020 [Available from: https://eidsvaag.no/.
6. BM B. Supply Chain design and analysis: Models and methods. Int J Production Economics. 1998:281-94.
7. BV. Bureau Veritas 2020 [Available from: https://www.marine-offshore.bureauveritas.co.uk/.
8. Guidelines for autonomous shipping - Guidance Note NI 641DT R01 E, (2019).
9. Bolbot V, Theotokatos G, Bujorianu LM, Boulougouris E, Vassalos D. Vulnerabilities and safety assurance methods in Cyber-Physical Systems: A comprehensive review. Reliability Engineering & System Safety. 2019;182:179-93.
10. Eloranta S, Whitehead A. Safety aspects of autonomous ships. In: Gl DNV, editor. 6th International Maritime Conference; Germany, Hamburg, 2016. p. 168-75.
11. Wingrove M. Shipborne systems most vulnerable to cyber-attack. Marine electronics & communications [Internet]. 2017; 11(3):[27 p.].
12. Bolbot V, Theotokatos G, Boulougouris E, Psarros G, Hamann R. A novel method for safety analysis of Cyber-Physical Systems - Application to a ship exhaust gas scrubber system. Safety. 2020.
13. Leveson N. Engineering a safer world: Systems thinking applied to safety. Moses J, de Neufville R, Heitor M, Granger M, Pate-Cornell E, Rouse W, editors. London, England: The MIT press; 2011. 560 p.
14. Bolbot V, Theotokatos G, Boulougouris E, Vassalos D. A novel cyber-risk assessment method for ship systems. Safety Science. 2020; Currently under review.

9. AUTHORS BIOGRAPHY

Victor Bolbot is a Research Associate at the Maritime Safety Research Centre (MSRC), Department of Naval Architecture, Ocean and Marine Engineering (NAOME), University of Strathclyde, Glasgow. His research focuses on safety and cybersecurity of marine autonomous and complex systems. His recent research output includes publications on safety analysis of power systems on cruise ships, scrubber systems, dual-fuel engines and cyber security risk assessments.

Gerasimos Theotokatos is the DNV GL Reader of Safety of Marine Systems at the Maritime Safety Research Centre (MSRC), Department of Naval Architecture Ocean and Marine Engineering (NAOME), University of Strathclyde, Glasgow. His research focuses on the development of scientific approaches to holistically capture the safety, energy and sustainability interplay of the complex marine systems including cyber-physical and autonomous systems by employing advanced model-based methods and tools for their design and optimisation pursuing life-cycle risk and energy management, efficiency improvement, and safety and sustainability enhancement.

Evangelos Boulougouris is the MSRC Director and RCCL Reader of Safety of Maritime Operations,

Department of NAOME, University of Strathclyde, Glasgow. His research focuses on the safety of ship operations, holistic design optimisation and design for safety. He is RINA Fellow and member of SNAME.

Lars Andreas Lien Wennersberg is a Research Scientist at SINTEF Ocean in Trondheim, Norway. He has a background from simulator-based testing of maritime control systems and engineering of maritime power, distribution and propulsion systems. His research focuses on design and test methods of autonomous ship systems.

Håvard Nordahl is a Research Scientist at SINTEF Ocean in Trondheim, Norway. He has a background from simulator based testing of maritime control systems and offshore modification projects. His research focuses on standardisation of simulations, design and cost evaluations of autonomous ship systems and autonomous shipping in general.

Ørnulf Jan Rødseth is a senior scientist at SINTEF Ocean and is the manager of Norwegian Forum for Autonomous Ships. He has more than 25 years' experience in maritime information and communication technology. In the last years, he has worked mainly with autonomous ship technology and maritime digitalization. He is a member of ISO TC8 and IEC TC80 and regularly meets at IMO as observer for ISO.

Jérôme Faivre is Smart Ships Rules Manager in the Development Department of BUREAU VERITAS Marine & Offshore Division. After 22 years as a naval senior surveyor, he has developed a large experience in Marine and Offshore fields and engineering management. His main responsibilities include development of Rules for autonomous and remote-controlled units.

Marco Molica Colella is Team Manager and Senior Innovation consultant in Italy at PNO Consultants. He has developed a large experience (+14) since his PhD as Engineer in the mechanical and aeronautics domain. He now supports the management, growth and finance of innovative projects in a large ecosystem made of primary actors in the energy and transport domain. He is the appointed AUTOSHIP coordinator.

